THE
Space Saver Book

THE
Space Saver Book

STEPHEN CORBETT

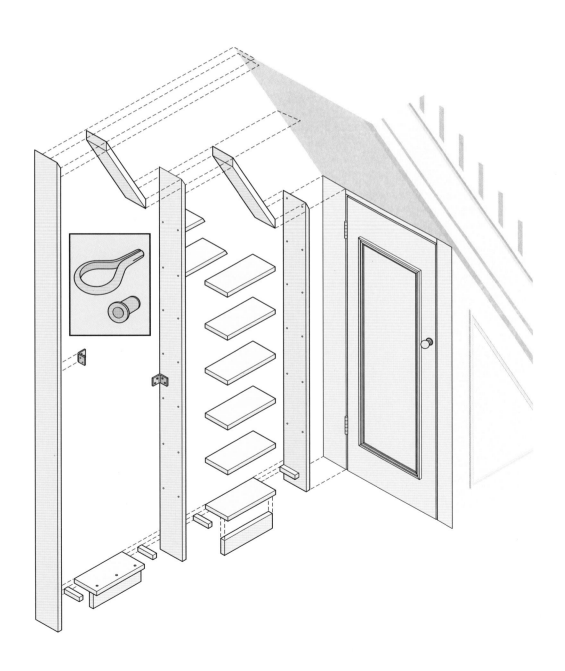

Abbeydale Press

Published in 1999 by
Abbeydale Press
An imprint of Bookmart Limited
Desford Road
Enderby
Leicester
LE9 5AD

© Bookmart 1999

Produced by Kingfisher Design, London N2 9NR

ISBN 1-86147-025-8

Project Director: Pedro Prá-Lopez
Technical Editor: Mike Trier
Technical Illustrator: Andrew Green
Supplementary text: Linda Doeser
Design and page formatting: Frances Prá-Lopez, Frank Landamore, Chris Brodrick
Project picture research: Stephen Corbett, Pedro Prá-Lopez
Supplementary picture research: Charlotte Deane, Katie Cox (Elizabeth Whiting Associates)
Index: Hilary Bird

Kingfisher Design would like to thank Elizabeth Whiting (Elizabeth Whiting Associates)
for her help and advice in the preparation of this book

Every effort has been made to trace copyright holders. Any unintentional
omissions or errors will be corrected in any future edition of this book

Colour reproduction by Global Colour Separation Ltd
Printed and bound in Italy

1 3 5 7 9 10 8 6 4 2

Contents

Introduction

Seeking to create order out of chaos is a defining human characteristic. Creating a well-ordered home is one of the first steps in adapting our surroundings to suit our individual taste and needs. A priority is to make use of every inch of space to provide storage for all our possessions in a way that is both practical and aesthetically pleasing. If you have ever looked in the window of an interior design store, or at a glossy lifestyle magazine, and thought to yourself 'I could make that – and at half the price, too' then this book is for you. Here we have selected a range of professionally built space-saving solutions, and modified them to produce a variety of projects tailor-made to match the needs of every practical home-owner.

Beyond the personal satisfaction of making something yourself, and the real financial saving this brings, you have a unique advantage in knowing better than anyone what is needed, and exactly how you want it to look. In this book we suggest ideas for eighteen different space-saving projects which you are free to adapt in any way that suits your taste. As you will be tailoring the finished project to fit into the space available, we have left the exact dimensions up to you. Similarly, the finish, whether paint, stain or varnish, will ultimately be your choice. Add the principles of good design and construction to be found in this book and you can benefit from the experience of a professional woodworker to achieve a really durable result. Furthermore, to help you make the most of the space-saving potential of each project, at the end of every chapter we have included a range of useful storage accessories which can be incorporated in the design.

There's nothing in this book that could not be tackled by anyone with a practical turn of mind and a keen interest in woodwork. Each project begins with the background information you need to start the job, followed by step-by-step diagrams for each stage of construction, with hints on special techniques. Look out for the ★ symbol which indicates a time-saving tip or a handy trick of the trade. With a selection of the tools illustrated opposite, and easy reference to the glossary where any unfamiliar materials or technical terms are clearly explained, you have everything you need. With the help of this book you can improve your home and your woodworking skills as you go along – just choose your first space-saving project, find a tape measure, and you're already on your way.

Stephen Corbett

STEPHEN CORBETT

BASIC TOOLKIT

Most of these items can already be found in the practical householder's toolbox. Any additional ones you need can be bought at relatively low cost from any DIY store or toolshop. With this basic toolkit you can tackle most of the projects in this book without hesitation.
Keep all tools, equipment and materials out of the reach of children.

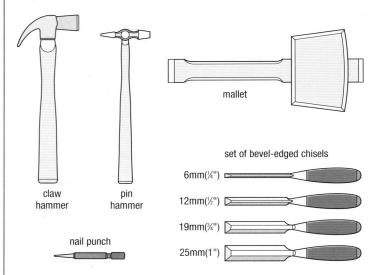

mallet

set of bevel-edged chisels

6mm(¼")
12mm(½")
19mm(¾")
25mm(1")

claw hammer

pin hammer

nail punch

EVERYDAY ESSENTIALS

Keep a selection of these basic essentials handy and they'll come in useful for the quick construction of most of the projects.

sanding block and sanding sheets

POWER TOOLS

Power tools are of NO USE AT ALL unless you:

- **FIRST** study the safety instructions
- **ALWAYS** observe those instructions
- **NEVER** forget that most accidents **only** happen to the careless user

Follow this code and they will make your woodworking safe and easy.

power drill
choose a dual-purpose model for woodworking and, with hammer action engaged, for drilling masonry

USEFUL EXTRAS

A few specialist tools used by the professional woodworker. Where the ✚ symbol appears in a project, you'll find these useful in adding speed and accuracy to your work, turning a tricky task into a simple one. Turn to the Glossary at the back of the book for explanation of these tools and their uses. Expensive items can always be hired very cheaply for a specific job if you have no permanent need for them.

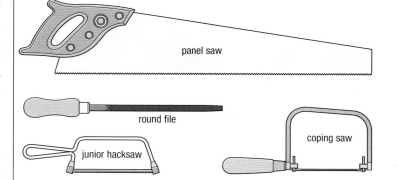

panel saw

round file

junior hacksaw

coping saw

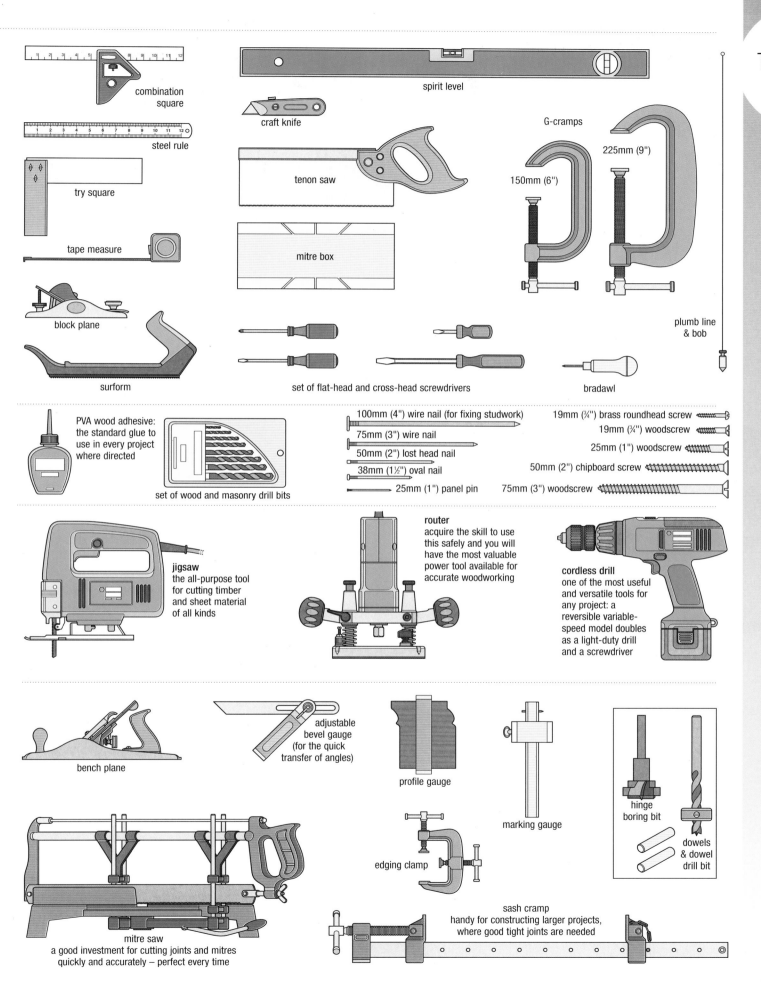

combination square

spirit level

craft knife

G-cramps

225mm (9")

150mm (6")

steel rule

try square

tenon saw

tape measure

mitre box

block plane

plumb line & bob

surform

set of flat-head and cross-head screwdrivers

bradawl

PVA wood adhesive: the standard glue to use in every project where directed

100mm (4") wire nail (for fixing studwork)

75mm (3") wire nail

50mm (2") lost head nail

38mm (1½") oval nail

25mm (1") panel pin

19mm (¾") brass roundhead screw

19mm (¾") woodscrew

25mm (1") woodscrew

50mm (2") chipboard screw

75mm (3") woodscrew

set of wood and masonry drill bits

jigsaw
the all-purpose tool for cutting timber and sheet material of all kinds

router
acquire the skill to use this safely and you will have the most valuable power tool available for accurate woodworking

cordless drill
one of the most useful and versatile tools for any project: a reversible variable-speed model doubles as a light-duty drill and a screwdriver

bench plane

adjustable bevel gauge (for the quick transfer of angles)

profile gauge

marking gauge

hinge boring bit

dowels & dowel drill bit

edging clamp

sash cramp
handy for constructing larger projects, where good tight joints are needed

mitre saw
a good investment for cutting joints and mitres quickly and accurately – perfect every time

CHAPTER 1

Children's rooms

Children grow up – literally as well as figuratively – astonishingly fast and the challenge with permanent storage in a child's room is to make it flexible enough to adapt. A built-in lowline wardrobe may be perfect for a toddler, but will be inadequate for a teenager, whereas shelves that once held rows of cuddly toys will be just as useful for books and CDs. A very important consideration is safety. Toddlers will heave themselves up by anything handy, so freestanding storage should be too heavy for them to pull over and anything built in, such as shelves, must be firmly attached and unable to tilt. Beware, too, of anything they can shut themselves or their fingers in. When children are young, storage that is fun is a good choice. For example, battens screwed to the wall and floor, with a triangular 'roof' frame and draped in brightly coloured fabric makes a tent cupboard. A boat- or car-shaped toy box doubles for storage and imaginary games. As they grow up and into their teens, they tend to treat their bedrooms as one-room apartments and need storage for a mixture of bedroom and living room items: lots of clothes and shoes, sports equipment and hobby materials, music centres, cassettes and CDs as well as computers and computer games.

PROJECT 1: *right.*
A bed on a raised platform with masses of space on wide shelves beneath, as well as a seating or play area, will delight children and may stay in use, with slight adaptations, as they grow older.

PROJECT 2: *below left.*
Easily accessible underbed storage can encourage a teenager to keep his room uncharacteristically tidy. Two cupboards and two drawers provide lots of space without encroaching on the rest of the room.

PROJECT 3: *below.*
A wooden chest with a hinged lid, painted with a colourful picture, makes a perfect toy box, especially when it has been designed with safety in mind.

Space-saving bed

There can be few children who would not relish the prospect of a bed designed and built especially for their own use, which serves as climbing frame, play area, storage for treasured possessions, and a space they can call their own. Safety of course is paramount for a raised bed platform, but you should not underestimate a small child's confidence or agility, which would put most adults to shame. With good design, and the use of robust materials, you can produce a unique article of bedroom furniture that will last for years. It can be adapted as the child grows up, to provide a study area, for example, by adding a small desk or workstation below the bed platform.

Although not illustrated in the photograph, this construction has been designed in such a way that a detachable rail can be inserted at the front of the bed platform for your child's safety, particularly in the early years of use. If you make the unit freestanding, and not built into an alcove as shown, you will need to provide a headboard or footboard to enclose the mattress completely – see steps 9 and 10 for details. You may also wish to consider fixing the unit to the rear wall for extra stability.

This clever design uses interlocking uprights and shelves, which make assembly a simple process. The

A fun space on two levels providing ample space for sleep and for play.

shelves at one end also serve as an integrated ladder with built-in hand-holds – note how they alternate to each side to make them easier to use. The two ends are then linked together at the top by the inner and outer rails to form the raised platform. When in future years the child outgrows the bed, the construction can be dismantled and the end supports used as freestanding shelf units in their own right.

MEASURING UP

It is important to design the size of the bed with safety in mind: the height should be such that a small child will have headroom underneath, and yet not so high that the unit will feel unstable.

- The shelves and uprights are made from substantial material 25mm (1") thick, which though heavy to handle during assembly will guarantee a solid and stable construction when complete
- The size of each step, and the interval between them, should be suitable for young arms and legs: no more than about 200mm (8") is suitable
- The height of the outer rails depends on the mattress: the top edge should be some 50mm (2") lower for comfort when seated on the edge of the bed
- Experiment with the height of the safety rail so that your child feels safe but not too enclosed: it is detachable so that making the bed is easier, and can be dispensed with altogether when the child grows older

TOOLS

- Basic Toolkit
- 25mm (1") chisel
- G-cramps
- Jigsaw
- Masonry drill and bits
- Pin hammer
- Round file ✪
- Router
- Sanding block
- Surform or block plane
- Try square

★ Router safety

This project requires the use of a router for making accurate joints. Always observe these safety tips:
- Ensure the grooving cutter is securely fitted
- Double check that the direction of travel is correct before starting
- Keep power leads clear of work

MATERIALS

- Plywood 25mm (1"): shelves, uprights
- Planed softwood: 150x25mm (6x1"): outer rails, headboard, footboard
- 100x25mm (4x1"): inner rails, slats, safety rail
- 75x75mm (3x3"): optional wall batten
- Wallplugs

Assembling the space-saving bed

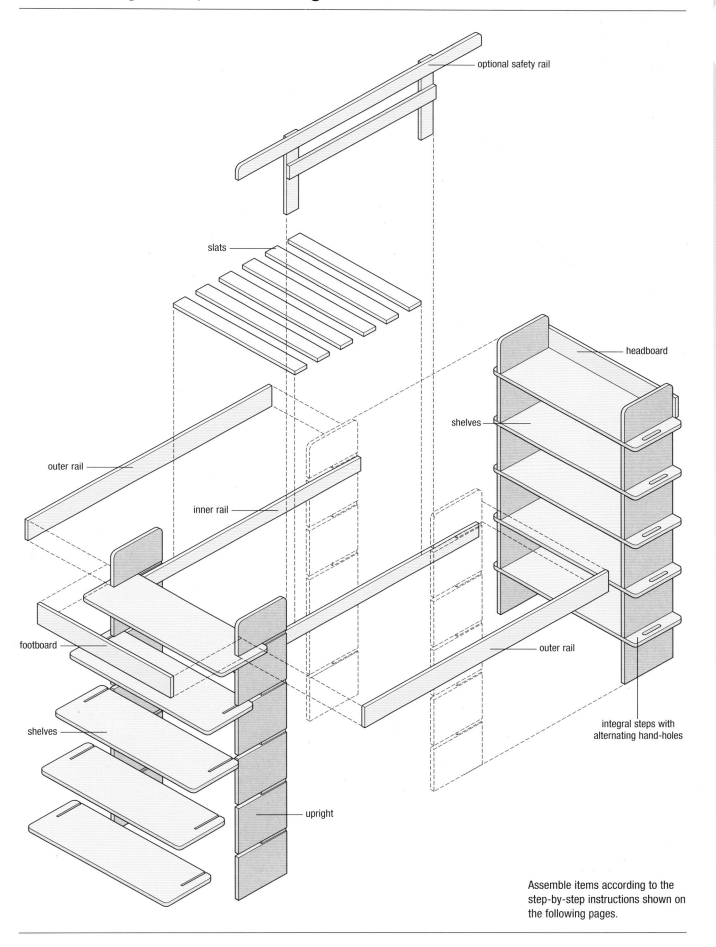

optional safety rail

slats

headboard

shelves

outer rail

inner rail

footboard

shelves

outer rail

integral steps with
alternating hand-holes

upright

Assemble items according to the
step-by-step instructions shown on
the following pages.

Constructing the units

The **modular** design of this unit makes construction a straightforward process by following the step-by-step procedures outlined below. Assemble the shelves and uprights for the end supports first, and set aside for glue to dry before completing the final stage by linking them together. The strength of the shelves, and the integrated access steps which form part of them, depends on a grooved **halving joint** for which a router is required. This is a useful item of equipment for the home woodworker, and is well worth acquiring, although it can also be hired at a reasonable cost specifically for the project. Take care to follow the instructions for the safe use of this powerful tool and you will find it a real asset.

When routing the grooves for the shelves, use a spare piece of wood to make a trial run, until the width and depth of the groove is satisfactory for a snug-fitting joint. Even a professional woodworker would take this precaution before grooving all the uprights. The other feature of this construction to which you should pay close attention is the quality of the finish. All exposed corners and edges should be rounded off with a plane or surform, followed by both coarse and then fine sandpaper, to avoid all risk of injury to the child. Fill the exposed end grain of all the plywood components with a proprietary wood-filler mixed to a smooth paste, then rub down and apply primer. Finish with non-toxic paint as recommended for children's toys or furniture – seek advice from the supplier if in any doubt.

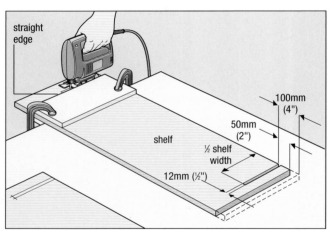

1 Begin by cutting the ten shelves to size: note that the five shelves at the head of the bed are 50mm (2") longer at the front to form the steps. Place all the shelves together and mark out for two 12mm (½") slots, adding 12mm (½") to the width of the mattress to determine the distance between them. Cut the slots carefully with a jigsaw, using a straight edge clamped to the work for accuracy.

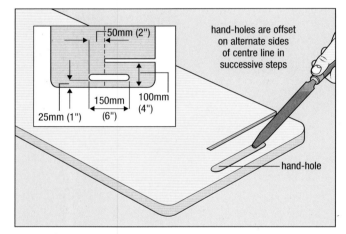

2 Form hand-holes in the steps by drilling two 25mm (1") holes 150mm (6") apart *(see inset)*, and remove waste with a jigsaw. Smooth the edges with a round file, or use coarse sandpaper rolled around a wooden dowel. For the safety of small children, round off the corners of each shelf to a 50mm (2") radius – ★ draw round a suitably sized paint tin as a quick way to avoid making a template.

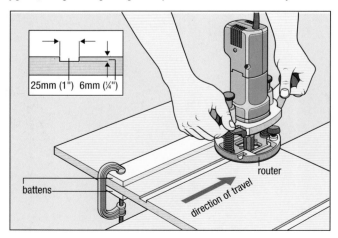

3 Cut the uprights to length and round off the top corners as in step 2. Mark out the shelf positions, and make the grooves which form the first part of the **housing joint**. Use a router fitted with a 25mm (1") cutter, set to a depth of 6mm (¼"). Clamp two straight-edged battens to both sides of the work to act as a guide, and always operate the router in the approved direction, as shown.

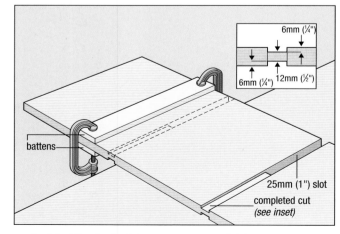

4 Adjust the depth of the router cutter to slightly over 12mm (½") and run over the groove again, but this time stop halfway across the upright, making a 25mm (1") slot. Turn the panel over and repeat the operation on the other face, making sure the second groove is exactly aligned with the first one as shown in the inset: check that the two battens are firmly clamped to the work.

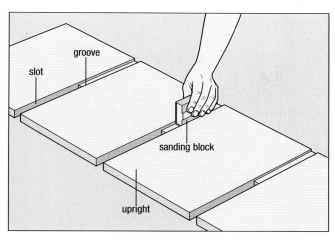

5 Repeat this procedure for all the uprights, and square off the end of each slot with a 25mm (1") chisel. Use a piece of sandpaper, wrapped around a block of wood just slightly thinner than the shelves, to clean up the grooves and ensure a good fit. You now have a combination of a **halving joint** and a **housing joint**, which affords a strong, simple construction when accurately made.

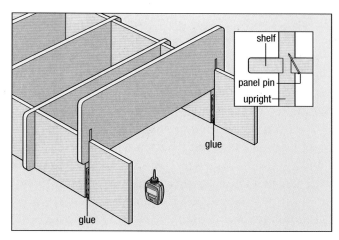

6 Lay the uprights on their back edges, apply adhesive to all sides of the grooves and slot the shelves into position as shown. Secure the joints with 30mm (1¼") panel pins, nailed at an angle from below, taking care to position the nail correctly to avoid splitting the plywood *(see inset)*. Check that the assembly is square, wipe off excess glue and leave to set before proceeding.

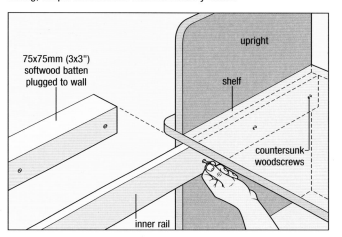

7 You can now erect the two units into position. They are linked together by the inner rails, fixed with countersunk screws from the inside. Although the unit is freestanding you may wish to secure it for extra safety. Screw through the inner rail into a length of 75mm (3") softwood, fixed to the rear wall with heavy duty wallplugs: this also serves to fill the gap at the back of the unit.

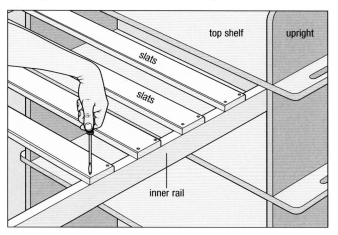

8 The two top shelves form the support for the ends of the bed: use slats of 100 x 25mm (4x1") planed softwood, which provide more flexibility and ventilation than solid plywood, to make the rest of the platform. Cut them to length, round off the edges, and fix to the inner rails with countersunk screws, making sure you leave no sharp edges which could damage the mattress.

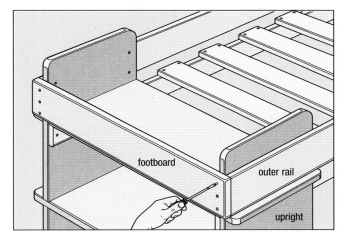

9 Fix the two outer rails to the top of the uprights. As well as locking the unit together securely, they also act as retainers for the mattress and bedding. Finally add the headboard and footboard as shown: make sure you screw into the uprights and not the rails, for greater rigidity. Again, make sure that the top edges especially are sanded smooth and free of splinters.

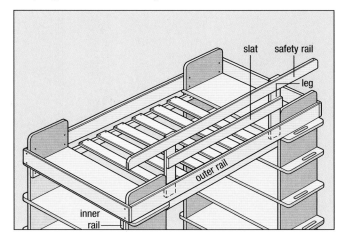

10 OPTION: For extra security, provide a removable safety rail which simply slots into position. Two lengths of 100x25mm (4x1") softwood form the legs, which being the same thickness as the uprights will slide down between the inner and outer rails of the bed platform. Fit a horizontal slat to locate on to the outer rail as shown, and add further slats to reach the required height.

Underbed storage

This space-saving bed is highly suitable for a teenager's room, where it seems no amount of storage is ever enough to keep the room tidy. Here by building a simple box bedframe you can make the entire space below accessible for storage of different items. This design shows two shallow drawers on one side of the bed only for clarity, but where there is access from both sides you can add more drawers as desired. By using telescopic runners, they can be pulled out almost to their full extent whilst still being securely supported. Similarly you can provide further shelves inside the large compartments, and vary the door sizes as you wish, but do bear in mind that the easier it is to access the storage cupboards, the more likely they are to be used.

This project is pure simplicity from start to finish, using basic techniques encountered elsewhere in this book. The construction of the bed itself is quite straightforward, using standard sized MDF boards with a minimum of cutting, and softwood slats as in the previous project. Be sure to **chamfer** all exposed

A tidy bedroom can become a reality with built-in underbed storage.

edges to prevent injury when walking around the room at night. The example shown in our picture has been made from veneered material as an alternative to a painted finish, but it would be advisable in this case to apply a coat of clear lacquer or varnish to make it more hard-wearing.

MEASURING UP

The size of your mattress will determine the overall dimensions of the bed, as the box construction encloses it fully. Allow an extra 12mm (½") all round if possible to assist with bed-making.
- Make the height of the front panel roughly equal to the thickness of the mattress: after fixing the top rail and the slats the top of the mattress will be a comfortable 50mm (2") above the top edge
- The telescopic drawer runners recommended are usually 500mm (20") long, so the maximum depth of drawer would be no more than 600mm (24") or half the width of a typical 4' bed

MATERIALS

- MDF 19mm (¾"): back and front panels, end panels, dividers, doors, drawer fronts
- Planed softwood: 100x25mm (4x1"): slats 50x50mm (2x2"): top rails 50x25mm (2x1"): bottom rail
- Plywood 12mm (½"): drawer boxes
- Softwood moulding 12x6mm (½x¼"): inset strips for doors and drawer fronts
- Drawer runners
- Sprung lay-on hinges
- Handles

TOOLS

- Basic Toolkit
- Block plane
- Drill and drill bits
- Jigsaw
- Screwdriver
- Try square
- **OPTIONAL:**
- Router

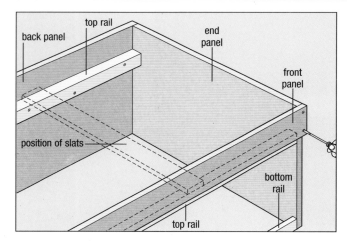

1 The bed frame is a simple three-sided box glued and screwed together. The front panel forms the fourth side, securing the mattress and allowing maximum storage space below. Cut the top and bottom rails to length and assemble the basic **carcass**, making sure it is square. Glue and screw the top and bottom rails in position and add the slats, spaced approximately 100mm (4") apart.

Assembling the underbed storage

Assemble items according to the step-by-step instructions shown on these pages.

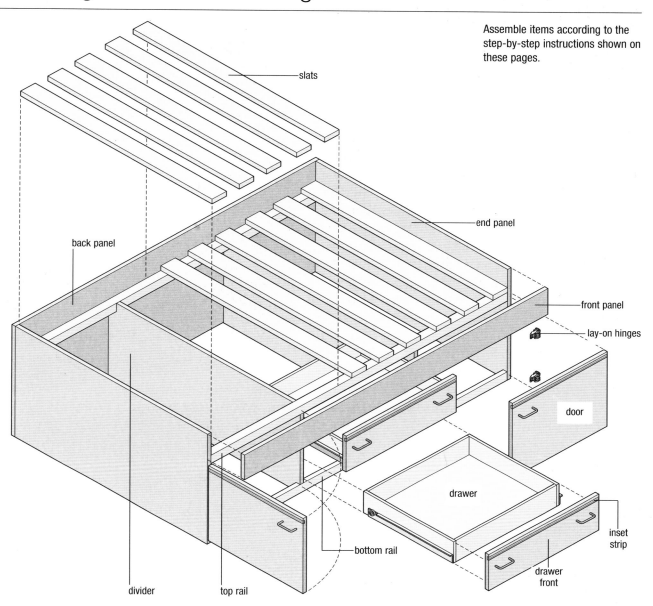

slats

back panel

end panel

front panel

lay-on hinges

door

drawer

inset strip

drawer front

bottom rail

divider

top rail

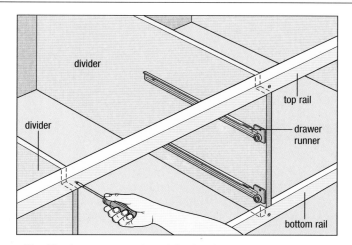

divider

divider

top rail

drawer runner

bottom rail

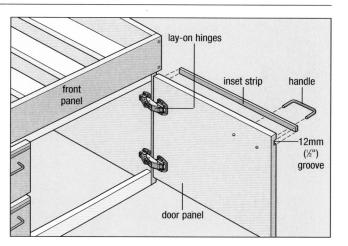

lay-on hinges

front panel

inset strip

handle

12mm (½") groove

door panel

2 The drawers are made up following the same procedure as for Fitted drawers, page 38. Cut two dividers, to the height of the top rails, and make notches for the rails with a jigsaw. Work out the position of the drawer runners and screw to the dividers for ease of assembly, then fix dividers in position as shown. Remember to allow 12mm (½") clearance for the runners at each side of the compartment.

3 Attach the two door panels with sprung lay-on hinges, *(see Understairs cupboard, page 77)*. These screw to the inside of the doors without any fixing holes, and need no catches to stay closed. For a final decorative detail, use a router as in the previous project to make a 12mm (½") groove in the doors and drawer fronts, then glue in an inset strip of wood painted to match the handles.

15

Painted toy chest

If you have young children of nursery age, a brightly coloured toy box is a simple project which serves a dual purpose: the child learns the value of a special place where all playthings can be put away, and you can attempt to keep the room reasonably tidy at the end of every day. Use it as an opportunity to decorate the nursery or playroom with a painting of your child's favourite character or theme.

Remember too that the toy box will be used for more than just storage – young children will climb on, around and inside it at the first opportunity, so observe a few safety points in the construction. Make cutouts in the sides for ventilation, and fit the recommended friction lid stays which prevent the top slamming down unexpectedly.

Countersink all woodscrews, and use round-headed screws for the fittings and piano hinge. Round off all sharp corners and edges before painting and always use lead-free, non-toxic paint, specially manufactured for children's furniture, available from toyshops as well as hardware stores.

Attractive, fun and convenient storage for toys at the end of the day.

MEASURING UP

Make the size of the box to suit your own needs, ensuring however that it is not too top-heavy, to avoid the risk of it being pulled over.

- If you want the box to double as a window seat, around 500-600mm (20-24") is a comfortable height
- Use a rounded object, such as a small tin of paint, to give a suitable radius for the rounded corners and cutouts on the front, top and sides, as in Space-saving bed, page 12, step 2
- Choose moulded skirting board to match that already in the bedroom to trim round the bottom edge of the box

MATERIALS

- MDF 19mm (¾"): top, sides, back, front, floor
- Softwood 25x25mm (1x1"): battens
- Moulded skirting board 100x19mm (4x¾"): bottom trim
- Brass piano hinge 19mm (¾"): for top
- Friction lid stays

★ Fitting lid stays

Fit the lid stays before painting the chest, then remove them. Refit them when the finish is dry. This makes the painting easier and prevents paint from getting on to the stay mechanism.

TOOLS

- Drill and drill bits
- Jigsaw
- Mitre box
- Nail punch
- Pin hammer
- Rasp or round file ⊕
- Screwdriver
- Tenon saw

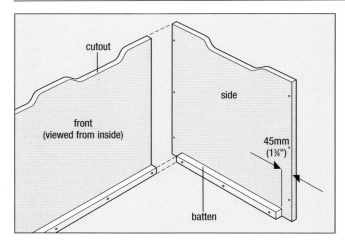

1 Prepare the front and side panels for assembly by making cutouts in the top edges as shown. Use a jigsaw to cut the shapes, then round off the edges with a rasp or coarse sandpaper. It is essential that all edges and corners have a smooth finish. Then screw battens flush with the bottom edges to all four sides – those on the side panels are 45mm (1¾") shorter at each end as shown.

Assembling the painted toy chest

Assemble items according to the step-by-step instructions shown on these pages.

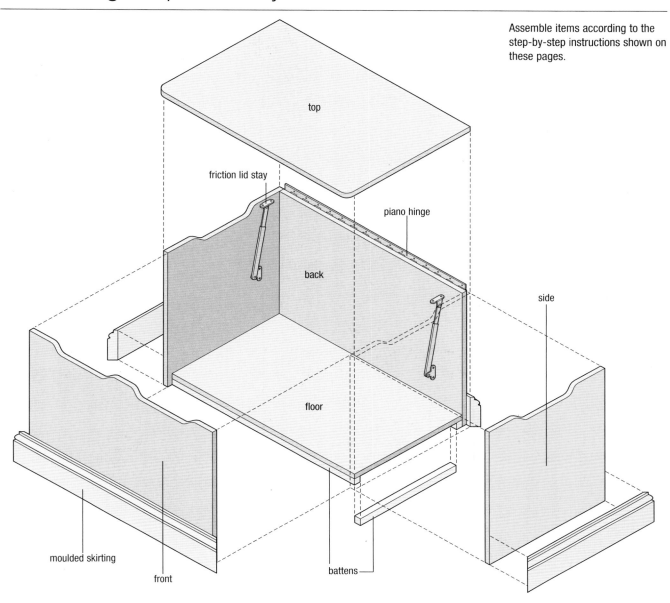

top

friction lid stay

piano hinge

back

side

floor

moulded skirting

front

battens

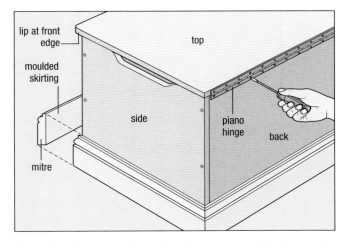

lip at front edge

top

moulded skirting

side

piano hinge

back

mitre

2 Screw the sides to the front and back and insert the floor, which will hold the box square and rigid. Cut the top to fit – note that it overlaps by 25mm (1") at the front edge to form a lip for ease of use. Fix with a strip of piano hinge to the back edge, as for the Pine linen box, page 41, step 3. **Mitre** four lengths of moulded skirting board to trim round the bottom and glue and pin in place.

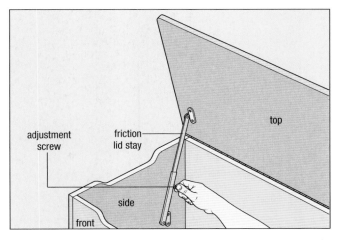

adjustment screw

friction lid stay

top

side

front

3 Fit the lid stays according to the manufacturer's instructions, and adjust the rate of closing with the friction adjustment screws. This is vital to prevent the top from slamming shut and causing harm to a small child. Sand the chest with fine sandpaper, then prime and undercoat. Apply a hand-painted or stencilled design, then finish all surfaces with two coats of clear, non-toxic varnish for child safety.

Instant storage ideas for children's rooms

Lightweight containers, such as plastic crates, wire mesh baskets, colourful cardboard boxes and even plastic buckets, bowls and rubbish bins are ideal for storing toys and other small items in a younger child's room, as they are easy and safe to handle. Alternatively, a hinged wooden box can do double duty as a toy box and a bench. A canvas laundry bag is a good way to store nappies.

A modular storage system that is easy to change and extend is ideal for a child's room, as it can be adapted as storage needs change. The baby's room has shelving for creams and lotions and a system of sliding baskets of different sizes for storing clothes and nappies.

The same system in a toddler's room now incorporates baskets for toys. The contents can be seen at a glance and they are lightweight and easy to clean.

The storage system is perfect for a playroom. Everything is immediately accessible but not dangerously under foot. A trolley with sliding baskets and a clip-on top makes a perfect child-sized table for painting or modelling and can be wiped clean afterwards.

The storage system continues to grow with the child and now a desktop with basket drawers beneath and shelves above transforms a child's bedroom into a teenager's study.

An inexpensive, folding pine bookshelf has been painted white and then decorated in primary colours by the child himself, providing a unique, but nonetheless useful, storage unit.

Shallow cardboard boxes covered in cheerful coloured paper are good containers for toys, building bricks and all sorts of clutter. They can be stored under the bed, leaving a clear play area.

This large dark blue plastic chest has lots of space for bulky toys. A matching painted pine chest of drawers keeps clothes tidy. The red metal chair can be folded flat to make more play space.

CHAPTER 2

Living rooms

One of the busiest and most visible places in the house, living rooms present particular storage problems. To be comfortable, they need to be tidy and orderly, but all manner of things must be easily to hand at different times of day, such as books, magazines, videos, casettes, CDs, drinks – in fact, all the paraphernalia of daily life. Shelves have a multitude of uses from storing books to displaying collectables. They can be easily built in an alcove or could extend from floor to ceiling along an entire wall, and there are many modular systems available in a wide variety of styles and materials. In a busy family house, where clutter tends to accumulate and the room has many different uses, cupboards may be a better choice, so that everything can be hidden behind closed doors. These can range from a simple cupboard built in an alcove, a third or half the height of the room, to a complete 'wall' of storage. A compromise, with some cupboards, some open shelves, solid doors and glazed doors, is probably the most versatile storage solution. It is also worth thinking about flexible storage in such a multi-purpose room. A cupboard with a bureau-style drop-down flap, for example, provides a place to keep writing materials and can be used as a desk, or it can serve to store glasses and bottles and as a table. Tables with shelves underneath keep magazines tidy and accessible.

PROJECT 1: *right.*
Shelving surrounds a walk-through doorway from living to dining room, creating an attractive architectural feature, as well as providing space for a surprising number of books.

PROJECT 2: *below left.*
Open shelves are easy and inexpensive to build in an alcove and can be used to display ornaments, as well as for more prosaic storage.

PROJECT 3: *below.*
This square coffee table, with a deep magazine shelf beneath, is both pleasing to the eye and pratical. As it is fitted with castors, it is easy to move without disturbing anything on it.

Walk-through shelving

A classical solution for storing books and displaying ornaments.

For an eye-catching way both to store and display books in the living room this ingenious set of shelves must be hard to beat. Even if your own home does not conform to the layout shown here the concept is a simple one which can lend itself to many situations. You may have a passageway connecting two rooms, or an L-shaped living room with a dining area off to one side. Alternatively, in a single large room this idea can be used to create a room-divider, making extra storage space where there was none before.

In this example a large archway connecting the living and dining rooms has been cleverly converted to make a doorway with storage housed on either side. One can imagine that on the far side the same shelving system might be used for the display of glassware or fine china. If the rather grand classical style in our picture doesn't agree with your own taste you can easily design an alternative which matches your home. Where there is no existing archway, for example in a passageway with a ceiling at full height, Step 1 overleaf demonstrates how to reduce the height of the ceiling by installing a simple partition to fit the overall proportions of the room, before building the unit itself.

Finish off the construction by choosing mouldings and skirting board to blend in with the rest of the living room, for the perfect built-in effect. The finishing touch of this project is the 'keystone', the dominant feature in the centre of the doorway. It is typical of such classical detail to combine decoration with structural function by locking the unit together and providing support for the high-level shelving.

MEASURING UP

The key to making a simple job of this project is to check first that the walls are truly vertical and parallel, which is not often the case. If this applies to you, take care to correct for this at the preparation stage (steps 1-3) to avoid problems later.

- If installing a high-level partition, make a prop to represent the height of the opening by nailing together two lengths of softwood, braced with an offcut of ply *(see step 1)*
- Measure the width of the aperture at its smallest point, subtract the thickness of the wall battens, and cut a **spacing rod** from a strip of plywood to help you in setting out the opening *(see step 2)*
- Measuring from the centre, mark out the width of the door opening on the spacing rod. Then add the thickness of the door casings. The remainder on each side will be the width of the shelving columns
- Select a suitable **architrave** of the right width to cover the wall battens at each side and to overlap just on to the edge of the shelving, as shown in step 9

TOOLS

- Basic Toolkit
- Adjustable bevel gauge ✚
- Coping saw ✚
- Jigsaw or panel saw
- Marking gauge ✚
- Masonry drill & drill bits
- Mitre box or mitre saw ✚
- 25mm (1") paring chisel
- Pin hammer and punch
- Plumb line & bob
- Profile gauge ✚
- Spirit level

MATERIALS

- MDF 12mm (½"): door casings
- Plywood 12mm (½"): shelves, shelf units, keystone, spacing rod
- Planed softwood: 100x50mm (4x2"): wall battens 50x50mm (2x2"): shelf battens
- Shelf moulding 25x12mm
- (1x½"): to trim shelves
- Architrave 100x19mm (4x¾"): to match existing mouldings
- Door moulding 25x19mm (1x¾"): to surround door casings
- Skirting boards: to match existing
- Rebated panel moulding: to surround keystone

Assembling the walk-through shelving

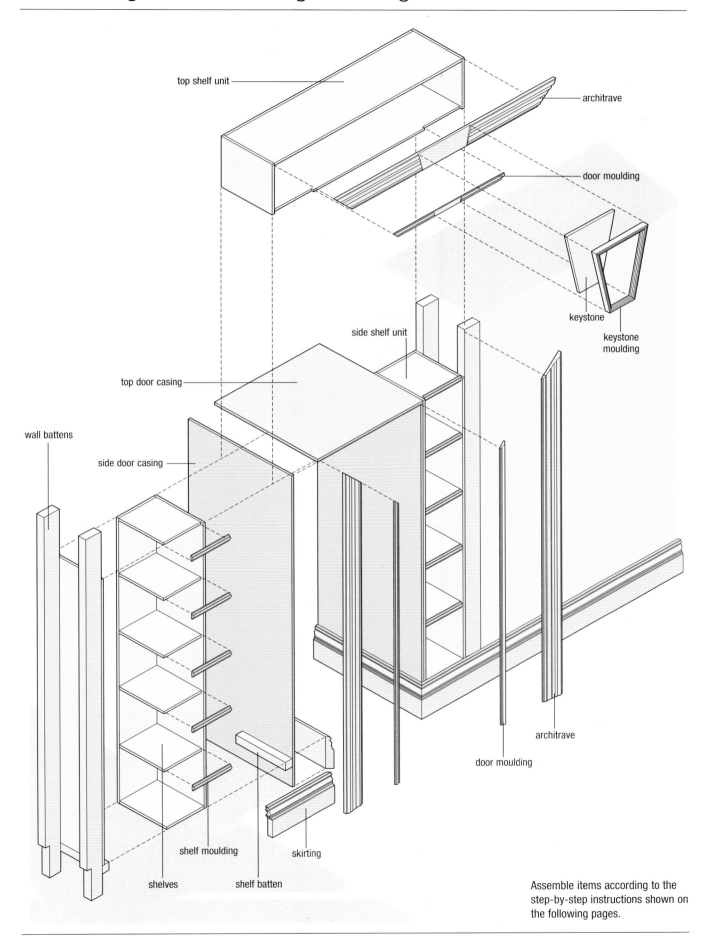

top shelf unit

architrave

door moulding

keystone

keystone
moulding

side shelf unit

top door casing

wall battens

side door casing

architrave

door moulding

shelf moulding

skirting

shelves

shelf batten

Assemble items according to the
step-by-step instructions shown on
the following pages.

Constructing the walk-through shelving

Although this construction may at first sight seem fairly complex, by breaking it down into a series of **modular** components it can be reduced to a simple process. Careful preparation is the first step: it is unlikely that the sides of the opening will be perfectly parallel and this can cause problems when it comes to sliding the units into position. It is easier to correct for any variation at an early stage, when fitting the wall battens, using the **spacing rod** at all times to achieve an aperture of constant width *(see Measuring Up)*. Any unsightly gaps where the battens meet the wall can be concealed later by fitting a suitable **architrave**.

The door casings are used to create the walk-through area, which can be any width, although you should try and achieve a standard doorway size if possible. Not only does this allow for the fitting of a door in future if you so decide, but you will find that most domestic furniture is designed to fit through a standard door opening of 760mm (30").

The shelving units can also be made to any convenient size, bearing in mind the books or other items which you intend to display. This is the great advantage of a purpose-made storage system. Note how the separate components are linked by using mouldings of the right size to lock the front edges together and provide a neat finish. Fill all joints and pinholes carefully, paint the unit to match the rest of the room, and the result will be an enviable addition which looks as though it has always been there.

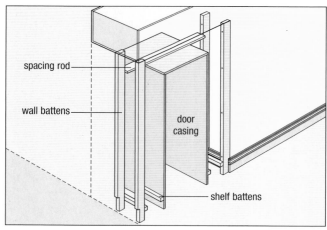

1 If you need to reduce the ceiling height in the opening, make up a simple **stud partition** at each end, using 100x50mm (4x2") softwood nailed together and fixed to the side walls. Make up a prop as a vertical spacer *(see Measuring Up)* and use a spirit level to ensure that the top of the opening is horizontal. Fix plasterboard or other wall cladding, fill holes and decorate to match the room.

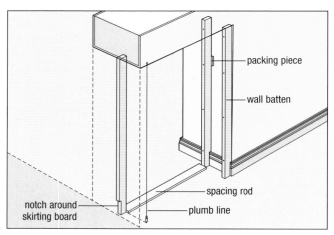

2 Prepare the side walls to suit the opening and the size of shelves and **architrave** you are using. Fix 100x50mm (4x2") softwood studs to the walls, using a plumb line and **spacing rod** *(see Measuring Up)* to create an aperture with true parallel sides. Use packing pieces of thin ply where the walls are out of true, and cut out notches if necessary to save removing the existing skirting board.

3 Cut the door casings to size, or have them pre-cut by the supplier to make sure they are square. Fix shelf battens to the casing sides, and to the wall battens, 12mm (½") below the height of the existing skirting board. Then screw the top to the sides and slide into place. Use your spacing rod, temporarily pinned across the top, to support the assembly centrally in the opening.

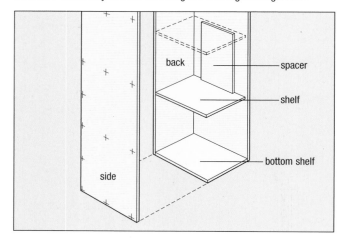

4 The shelves at each side of the unit are two simple columns. Note that the bottom shelf in each case is full depth, flush with the front edges of the side pieces, whereas all the other shelves are recessed by 12mm (½") to allow for the shelf moulding. Cut a square of plywood to use as a spacer for positioning the shelves, glue and pin them in place, then fix the back panel to hold the unit square.

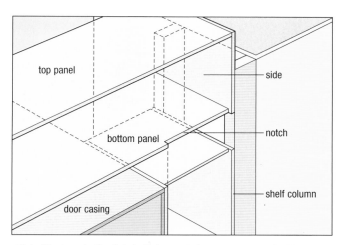

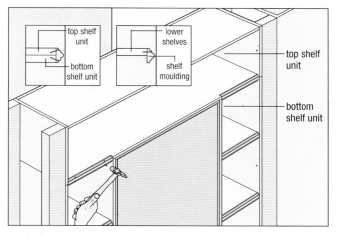

5 The top shelf unit is built in much the same way – double check the height remaining before cutting the sides. Before assembly, slide the bottom panel of the shelf unit into place, centralise it on the door casing, and mark the position of the casing sides. Add 12mm (½") each side, and cut two notches 12mm (½") deep, as shown, to match up with the top shelf on each side column.

6 Slide the shelf units into position, and screw to the wall battens and door casing. If you have done your preparation carefully all front edges should be flush and vertical. You can now add the shelf mouldings to trim the front edges – note how the moulding is used to link together the top and bottom shelf assemblies, with two rows of panel pins, for a neat and seamless finish *(see inset)*.

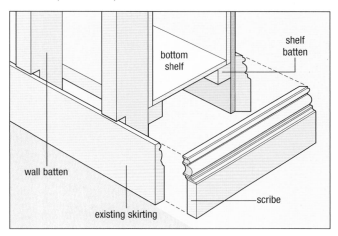

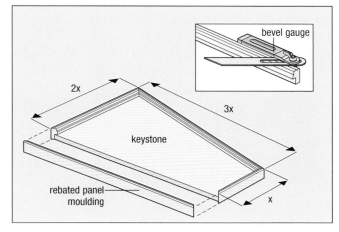

7 At the bottom of the unit the shelves are now resting on the lower shelf battens and should be flush with the top of the skirting board. Add new skirting, of the same height and pattern, to trim round the bottom of the door casings and the front of the bottom shelves. Refer to Fitted wardrobe, page 37, step 7 for guidance on **scribing** the skirting to the existing moulding with a **profile gauge**.

8 To make the keystone, cut a panel of 12mm (½") plywood to the proportions shown here, and trim with **rebated panel moulding**, glued and pinned in place. Use an adjustable bevel gauge to measure the angle at each corner, halve the angle, and make angled cuts at each end of the moulding to suit. ★ Look up Understairs storage units, page 73, step 8 for a handy tip on bisecting angles.

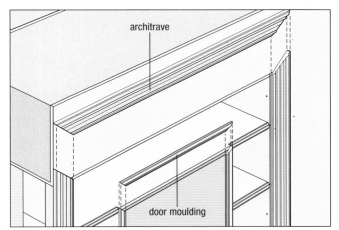

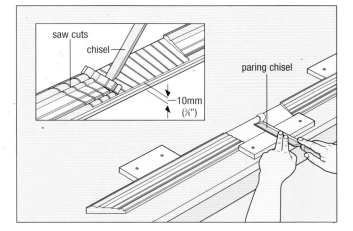

9 Cut the vertical architrave sections, **mitre** the top corners in a mitre box and pin in place. Then mitre the ends of the top piece but just rest it in position for the moment without fixing. Repeat the procedure with the door moulding which covers the inner edges of the shelves and the door casings. Offer up the keystone, score along the edges where they intersect the horizontal mouldings, and remove.

10 To make the **housing joints** in the mouldings, score a line 10mm (⅜") from the back face with a marking gauge, and make a series of vertical sawcuts down to this depth with a fine tenon saw to assist in chiselling away the waste *(see inset)*. Clamp the moulding with 10mm (⅜") blocks of wood screwed to the bench as shown, and pare the bottom of the joint flush for the keystone.

Alcove shelving

Shelving in the living room should be able to offer more than just useful storage space: it should also be attractive enough to work as an area for display or decoration. Although there is no shortage of ready-made shelving to choose from, a set of custom-built, well-fitted shelves makes all the difference. Whether they fill the entire alcove, or fit above a set of cupboards as in our picture, is up to you. If you want to make a matching set of base units, turn to Fitted vanity unit, page 46 to give you ideas.

A typical layout in many houses may feature a chimney breast with alcoves on each side, as in our picture. Fitting shelves to such areas should be a simple job but often turns out to present unforeseen problems. In our example the task is made even more of a challenge where the walls meet at an angle. Even a seemingly square recess proves to be otherwise when you come to install a ready-made bookshelf. By following the simple tips on measurement shown here you can build a set of accurately fitted shelves in just about any available corner of the room.

Simple and unfussy alcove shelving complements any interior.

MEASURING UP

The internal width of an alcove is always difficult to determine with an ordinary tape measure which is not the most accurate method for this procedure. The solution is to use two 'pinch rods' as illustrated in step 1 on the right.

- Cut two 25x19mm (1x¾") battens ensuring they are straight and free of any twist
- Make sure the two battens, when overlapped as shown, are long enough to fit the alcove at its maximum width
- Keep them horizontal and slide them alongside each other until they touch the opposing corners of the recess
- Then 'pinch' the battens together and use their combined length to transfer the exact measurement required
- Follow this tip for each shelf in turn as it is quite likely that the width of the alcove will vary from top to bottom

MATERIALS

- MDF 19mm (¾"):
  shelves
- Planed softwood
  50x25mm (2x1"):
  shelf trim
  25x19mm (1x¾"):
  wall battens

TOOLS

- Basic Toolkit
- Adjustable bevel gauge ⊕
- Jigsaw or panel saw
- Masonry drill and drill bits
- Plumb line & bob
- Screwdriver
- Spirit level
- Tenon saw

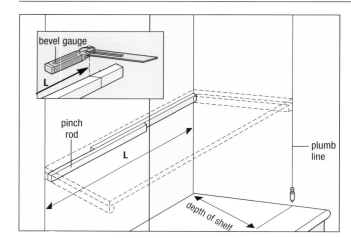

1 Use the pinch rods as described in Measuring Up to measure the width at the back of the alcove, and transfer the internal length L to the wall batten. Set the angle required with a bevel gauge *(see inset)* and cut to length. Mark the position of the front corners of each shelf with a plumbline, as shown, then measure and cut the side battens, using the bevel gauge again to achieve the correct angle.

Assembling the alcove shelving

Assemble items according to the step-by-step instructions shown on these pages.

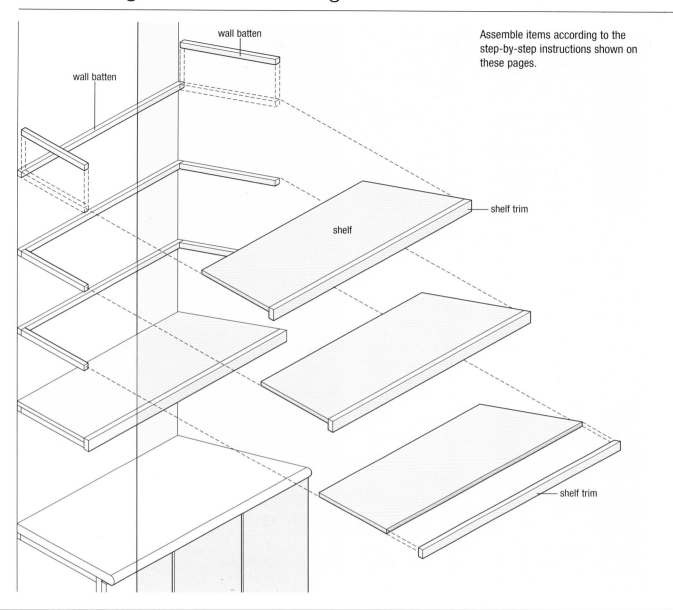

wall batten

wall batten

shelf trim

shelf

shelf trim

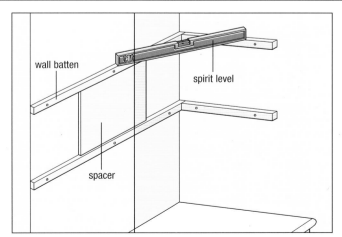

wall batten

spirit level

spacer

2 Fix the wall battens with suitable masonry or plasterboard plugs at approximately 300mm (12") intervals. Take care to keep the battens level in both directions. When the rear batten is fixed, double-check the sides by resting a spirit level at an angle across the corner, as shown. After fixing the battens for the lowest shelf, use a square offcut of ply as a spacer to set the height for the others.

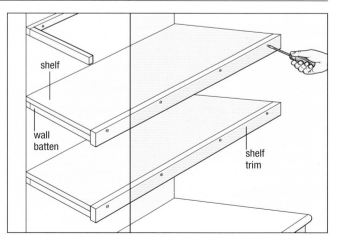

shelf

wall batten

shelf trim

3 Cutting the shelves to fit is now a simple procedure, using the pinch rods to measure the length at the back and front edges, and the bevel gauge to determine the angles at each end. Finally glue and screw the shelf trim battens to the front edge of each shelf – these not only make a tidy job by concealing the wall battens from view, but serve to strengthen the shelves and prevent sagging.

Coffee table on wheels

A low central table in the living room can sometimes be useful, but at other times turns out to be something of an obstacle. This neat little project solves that problem with a design for a low wheeled table, with a large surface area and generous storage space underneath for books, magazines or newspapers. Made from inexpensive materials, and of simple construction, it forms an attractive focal point for the living room, and can always be moved into a convenient corner.

This table has been made using white melamine-coated chipboard, providing a smooth surface that is easy to wipe clean. Any suitable sheet material could be used, however, depending on your choice of finish. Whatever the choice, the top is best finished off using a solid wood trim all around the edges, to conceal the core of the material from view and strengthen the whole construction. Use the same wood for the uprights and corner posts and you have a very neat design, integrally strong, using the minimum of components.

This simple, movable coffee table will enhance any modern living room.

MEASURING UP

The table top obviously does not have to be a perfect square, but can be any shape to suit your own living room. Even so, the construction method is identical.

- The top, base and divider can all be cut to the same front-to-back dimension from a single sheet of material, to ensure accurate assembly. Some suppliers equipped with a cutting department will do this for you at no extra charge
- Choose smooth-running swivel type castors for maximum manoevrability, of the right height to be just concealed by the overhang of the bottom edging (see main diagram opposite)

MATERIALS

- MDF or chipboard 19mm (¾"), melamine-coated: top, base, divider
- Planed softwood 38x25mm (1½x1"): top edging, bottom edging, uprights
- Castors, plate fixing, to suit

★ **Positioning castors**
Ball castors (shown opposite) are made in left- and right-handed versions; a set comprises two of each. To ensure easy movement, arrange them alternately round the base:

TOOLS

- Basic Toolkit
- Dowel drilling kit ⊕ (see step 2)
- Drill and drill bits
- Jigsaw or panel saw
- Sash cramps (if available) ⊕
- Screwdriver
- Tenon saw and mitre box

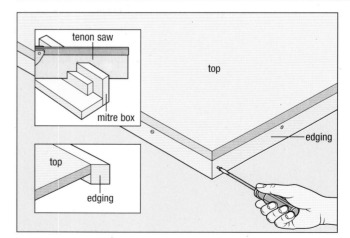

1 The top and base are identical in size and construction, being trimmed all round with softwood edging strips. **Mitre** the corners neatly in a mitre box, glue and screw all round the edges and set aside to dry. Note how the edging is flush at the top edge on both panels, and overhangs to form a lip at the bottom (see inset). Use narrow gauge screws to strengthen the mitres at each corner.

Assembling the coffee table on wheels

Assemble items according to the step-by-step instructions shown on these pages.

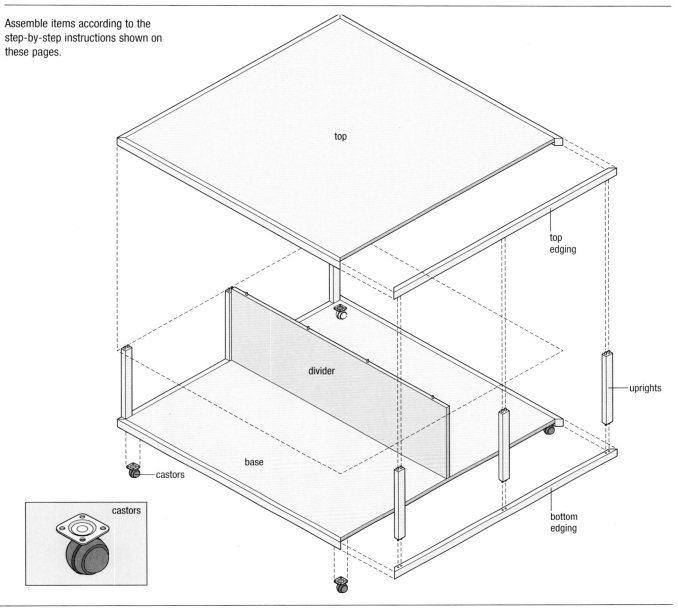

top

top edging

uprights

divider

base

castors

bottom edging

castors

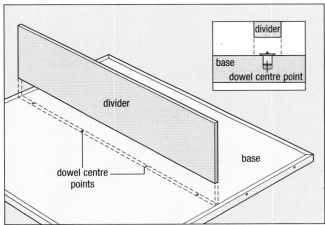

divider

base

dowel centre point

divider

base

dowel centre points

base

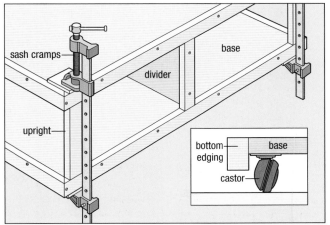

sash cramps

base

divider

upright

bottom edging

base

castor

2 The divider is connected with **dowel joints** to the top and base, not only for strength but also to make assembly easier. **Dowel centre points** *(see inset)* are available in kit form with the drill bit and dowels and make accurate location a simple matter – drill holes in the base, insert centre points and press the bottom edge of the divider on to the points to mark the positions of the matching holes.

3 The uprights are located with dowels in the same way: drill the holes as shown on page 40, step1. Accurate, well-glued dowel joints require no further fixing, avoiding any unsightly screws through the top. Use sash cramps to clamp the assembly together for really tight joints, or leave to dry under heavy weights. Finally fix a castor at each corner, ensuring that it can swivel freely.

29

Instant storage ideas for living rooms

Items of all shapes and sizes need to be accommodated in a living room and there is a vast range of purpose-built storage available, from simple stacking drawers for videos, through coffee tables with drawers or shelves, to modules for a complete home entertainment centre. Styles and materials range from the traditional, through the contemporary, to the outright zany.

A colourful plastic wine rack holds six bottles and will fit neatly into an available space vertically or horizontally.

A fun 50s retro style magazine rack, made of plywood, is both versatile and practical. It also provides space for paperback books and has two small shelves.

Simple, pale wood bookshelves are available in a range of sizes. Paper-covered filing boxes in bright primary colours are ideal for storing easily mislaid everyday items from bills to cotton reels. An easy-to-store pale wood folding chair is the perfect solution for people with small living rooms who like entertaining guests.

Stack CDs safely in a tower of interconnecting modules, each one a different colour.

A small reproduction miltary-style chest of drawers with a mahogany veneer and brass handles turns storage into art.

This casette rack overcomes the perennial problem of never being able to find the tape you want and then all those little plastic boxes falling everywhere when you do.

This stack of versatile self-assembly units, comprising two drawers and two cupboards, provides mobile storage on castors.

A director's chair with a leather seat and back rest provides comfortable seating, but can be folded flat and stored in quite a narrow space when it is not in use.

CHAPTER 3

Bedrooms

The bedroom should be a haven of peace and tranquillity, but can often become quite chaotic in the struggle to find sufficient storage space. Free-standing storage units often look haphazard and can take up an unnecessary amount of space, especially in a small room. A fairly inexpensive and easy solution would be to attach floor and ceiling tracks in an alcove and fit sliding doors. If these are then decorated to match the walls, they will hardly be noticeable, but clothes can be hung tidily out of sight. Alternatively, a practical and colourful solution would be to fit floor-to-ceiling shelving with open compartments for stacks of shirts and sweatshirts or shoes, making them visible and accessible. Baskets stacked on open shelves are an easy way to store smaller items, such as socks, or baskets with handles can be hung from two or three rows of hooks on the back of the bedroom door. Drawer dividers made from plywood and quadrant beading would keep unruly ties and belts in order. However, the ideal and most comprehensive approach for many people is to install fitted units which make the maximum use of the space, can be customized for specific needs and have clean uncluttered lines.

PROJECT 1: *right.*
An impressive fitted wardrobe with mirror panelling on the doors perfectly complements the elegance of this spacious bedroom. The design is flexible and can be easily adapted for a smaller room or one with a lower ceiling.

PROJECT 2: *below left.*
The best of both worlds – a self-assembly unit is adapted with strong drawers that can withstand the wear and tear of constant use.

PROJECT 3: *below.*
This dual purpose pine linen box opens at the front, so the top can be permanently in use as a table.

Fitted wardrobe

The master bedroom is one area in the house where you could be forgiven for permitting yourself a touch of luxury. There is no shortage of companies who will offer you a quotation for an expensive range of wardrobes and fitted storage units for the bedroom: you will be pleased to know that much the same effect can be achieved by the competent home woodworker, using the same methods as the professionals at a fraction of the cost.

The project we have chosen here uses the trick of mirrored doors to add distinction to a simple wardrobe unit. In a spacious room with high ceilings such as this one, the use of mirrors enhances the simple uncluttered layout; in a smaller bedroom you can use them to even more striking effect to add light and space. Buying the mirror glass for the doors will add some expense to the project, but you can save time and money by using ready-made doors, obtainable in convenient standard sizes, and fitting the mirrors yourself.

This design can be adapted to fit almost any bedroom: if the space available does not allow you to cover a whole wall, you may have a suitable alcove that is not being fully used. Alternatively, one of the side panels can be modified to form the end of a run of units if you don't wish to fill the entire wall. The

This wardrobe is capacious, while making the room appear larger too.

height can also be altered to suit your requirements: either build the unit full height to the ceiling, or fit a filler panel above, which can be decorated to blend in with the rest of the room decoration, as in our picture. Fit an optional top panel (see step 3) if required: the top cupboards should not be so high that they are of no practical use.

MEASURING UP

Decide on the ideal width of the unit, and try and select ready made doors to suit. A typical door size would be 1830x500mm (72x20"), and 400mm (16") high for the top cupboards.

- Remember to calculate the combined width of the frame uprights and subtract from the overall space available to determine the size of the door openings
- If you can't avoid a small discrepancy at one end, fix a vertical batten to the side panel and insert a narrow strip of plywood or plasterboard to fill the gap
- The depth of the wardrobe should be no less than 500mm (20") to accommodate a standard size coat hanger, and up to 600mm (24") would be more advisable to avoid crushing the contents
- Note that this design has no need for a floor panel inside the wardrobe: this provides a generous internal dimension for full height hanging, or alternatively extra storage space for shoes or other small items

TOOLS

- Basic Toolkit
- Block plane
- Coping saw ✛
- Jigsaw
- Marking gauge ✛
- Mitre saw for mouldings ✛
- Profile gauge ✛
- Spirit level

MATERIALS

- MDF 19mm (¾"): side panels, central divider, shelves, top panel (if required)
- Planed softwood 50x50mm (2x2"): frame uprights 50x25mm (2x1"): frame cross rails 25x25mm (1x1"): fixing battens, mounting blocks
- Ready made door panels, 19-25mm (¾-1") thickness
- Mirror glass 3mm (⅛"): to fit door sizes
- Rebated panel moulding 32x12mm (1¼x½"): for fixing mirrors
- Glazing tape
- Skirting board: to match existing skirting
- Magnetic touch latches: for upper cupboard doors
- Magnetic wardrobe catches: for lower wardrobe doors

Assembling the fitted wardrobe

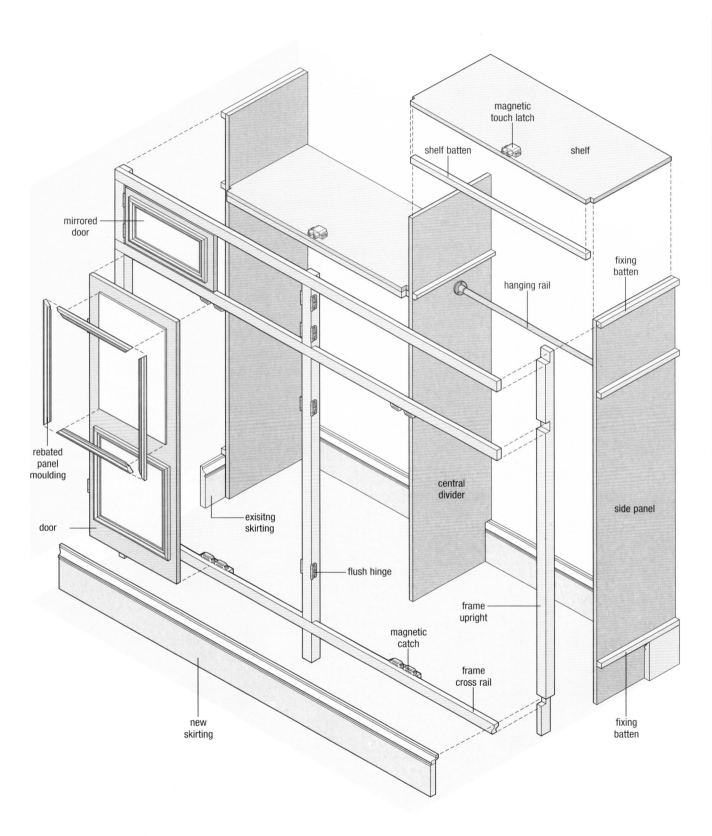

mirrored
door

magnetic
touch latch

shelf batten

shelf

fixing
batten

hanging rail

rebated
panel
moulding

door

exisitng
skirting

central
divider

side panel

flush hinge

frame
upright

magnetic
catch

frame
cross rail

new
skirting

fixing
batten

Assemble items according to the
step-by-step instructions shown on
the following pages.

Constructing the fitted wardrobe

The side panels, central divider and shelves form the basic **carcass** of this unit. Most DIY suppliers will cut MDF boards to the size required without extra charge, thus saving you the task of handling large sheets of material at home. Do your measuring up before going to collect your material, and prepare a cutting list of the main components.

Once the basic carcass is erected, the next step is to make the softwood frame, using simple **housing joints** to fit the cross rails to the uprights. Accuracy is essential here as this frame supports the doors, so check all the frame members are square and level as you proceed: here you can make up for any discrepancies in the carcass if the walls of the room are not perfectly vertical – they usually aren't.

Mirrored doors will be quite heavy, and it goes without saying that the mirrors should be securely fixed using the specially shaped mouldings shown here, but don't make the mistake of fixing them too tightly, to avoid cracking the glass. Use **glazing tape** (available from your local glazier), or self-adhesive foam strip, to cushion the glass, and fix the mouldings with decorative screws, to allow removal should the worst happen.

Notice that the upper cupboard doors in our picture have no visible handles. They use magnetic touch latches, which open the door automatically in response to a light pressure. This system is especially suitable for high level cupboards, and affords a smooth run of doors without handles of any kind.

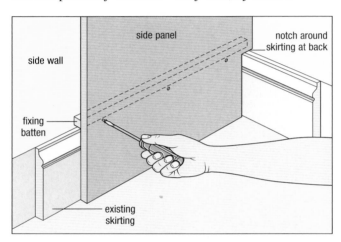

1 Having decided on the overall dimensions of the unit, the first step is to fix the side panels. If there are skirting boards around the room, fix battens to the side walls to set the panels clear of the skirting. Fix one just above skirting level, one at shelf height, and one at the top of the unit. At the back, notch the panels to fit around the skirting board, then screw them in position.

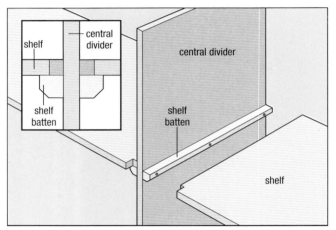

2 The central divider can be easily positioned by cutting the shelves first to a fixed width and using them as spacers. Fix battens to the sides and divider for the shelves, ensuring they are level. As with all battens which will be visible, a neat finishing touch is to **chamfer** the edges with a block plane before fixing, to remove any sharp or unsightly edges *(see inset)*.

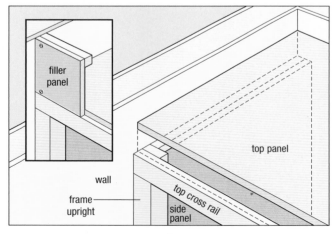

3 If your wardrobe does not reach all the way to the ceiling, and you are fitting a filler panel to make up the difference in height, you will need a top panel for the unit. It should overlap the sides and central divider at the front: cut it to the suit the overall depth of the unit, less the thickness of the filler panel, which should be flush with the top cross rail when fitted.

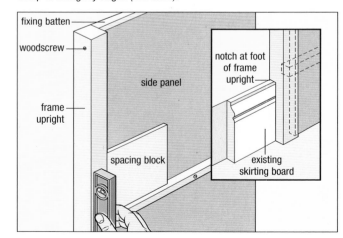

4 Cut the frame uprights to length, and fix temporarily in place with 75mm (3") woodscrews at top and bottom. Use a spirit level to ensure they are vertical, or the doors will not hang correctly: a spacing block will help to maintain the correct offset from the edge of the side panels. Cut a notch in the foot of each upright to fit around the skirting if necessary *(see inset)*.

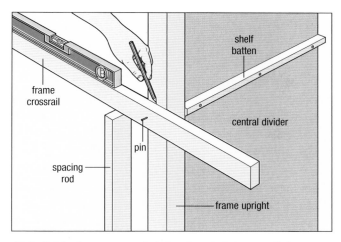

5 Cut the frame cross rails to length and pin temporarily in position ensuring they are level. Fit the bottom rail first, **scribed** around the skirting board at each end if required *(see step 7)*; then use two **spacing rods** the same height as your doors to locate the middle and top rails. With a sharp pencil mark out for the **housing joints** in the frame uprights.

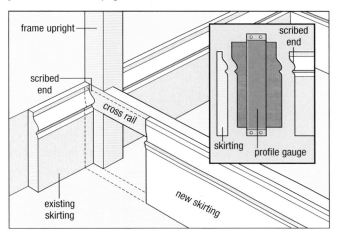

7 Glue and pin the cross rails in the prepared notches, and add a length of skirting board along the bottom rail to complete the unit. You may need to **scribe** the ends to fit existing skirting: use a **profile gauge** to transfer the shape of the moulding *(see inset)*, and cut away the waste with a jigsaw or coping saw. Finally fix shelf battens to the centre cross rail, 19mm (¾") from the top edge, and fit the shelves.

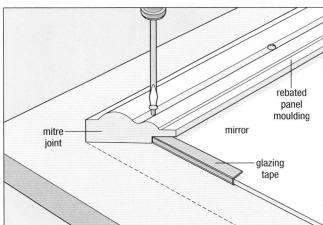

9 The mirrors are secured by **rebated panel mouldings**. Protect the mirror edges with **glazing tape**, which helps to cushion the glass. Select a moulding where the depth of the rebate allows for this. With the door lying securely on a flat surface, **mitre** the corners of the mouldings and fix with small brass round-head screws. For extra safety, use glue on the bottom moulding which supports the weight.

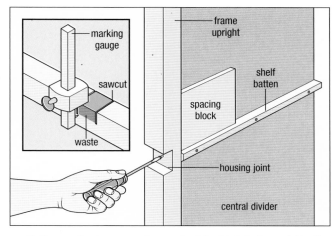

6 Remove the frame members and cut the joints in the uprights. Use a marking gauge to mark the depth of the cross rails, cut down to the line with a tenon saw, and remove waste with a sharp chisel *(see inset)*. Then glue and screw the uprights in position. Screw though the notched joints to reduce the number of visible screw holes. Where concealment is not possible, countersink the screws and apply filler.

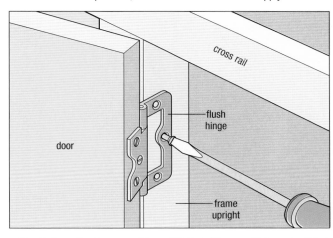

8 To hang the doors use 75mm (3") flush hinges, which can be fitted flush to the door and frame without a **rebate**: screw the small inner leaves to the door first, then offer up to mark the hinge positions on the uprights. Fit with just one screw and check the doors for a snug fit, allowing 1.5mm (¹⁄₁₆") clearance all round. Make any necessary adjustments before fixing the mirrors, while the doors are easier to handle.

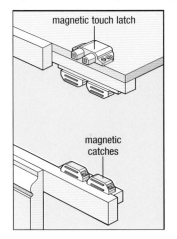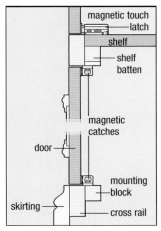

10 Hang the doors and fit doorknobs and catches. The upper doors may be secured with magnetic touch latches fixed to the shelves; the larger wardrobe doors are best fitted with heavy duty contact catches. At the top they can be fixed underneath the shelf battens; at the bottom, screw a small length of 25x25mm (1x1") timber to the lower cross rail as a mounting block.

Fitted drawers

The most attractive feature of this dressing table is the clean and simple design of the drawer fronts. The DIY alternative to a professional installation like the one in our picture is to fit self-assembly units: sadly the quality of the drawers supplied in such units often leaves a lot to be desired: they can wear out through wear and tear long before the carcasses themselves.

In this project we show how to make a set of long-lasting drawers, which won't stick or fall apart. The secret is to use good quality, low friction drawer runners. Being self-aligning and fully adjustable they make it easy to achieve flush-fitting drawer fronts.

You can have the best of both worlds, therefore, by installing self-assembly units in the layout of your choice, and making your own drawers and drawer fronts. We have added a timber frame surround to disguise the box-like appearance of the carcasses. For a really stylish finish you may wish to add a glass top to the unit. Use **laminated glass** for safety, and get your glazier to grind and polish the edges to reduce the risk of chipping.

Near-frictionless drawers with solid, traditional construction.

MEASURING UP

Divide the internal height of the **carcass** to suit the number of drawers. Decide on the height of each drawer front for good visual effect as well as a practical range of sizes.

- Make the drawer boxes separately for ease of construction and added strength. They are less deep than the drawer fronts themselves to allow for clearance *(see step 1)*
- If fitting a framed surround to the front of the units, fit spacing strips to the carcass sides to pack them out to the same thickness *(see step 2)*
- Measure the internal width of each carcass and subtract 12mm (½") each side to allow for the drawer runners
- Make a hardboard template for the top: use it to make the top panel, and take it to the glazier for cutting the glass

MATERIALS

- MDF 19mm (¾"): drawer fronts, top panel
- Plywood:
  12mm (½"): drawer boxes, spacing strips
  6mm (¼"): drawer bottoms, spacing strips
- Softwood:
  50x19mm (2x¾"): top rail

- 32x19mm (1¼x¾"): frame uprights
  25x25mm (1x1"): fixing battens
- Hardboard 3mm (⅛"): template for top
- Laminated glass: 6mm (¼") glass top
- Drawer runners & handles

TOOLS

- Basic Toolkit
- Panel saw or jigsaw
- Pin hammer & nail punch

OPTIONAL:
- Router with 6mm (¼") grooving cutter

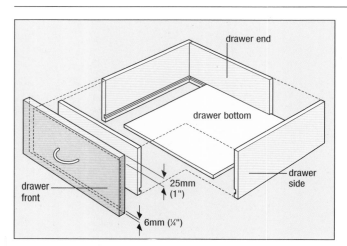

drawer end

drawer bottom

drawer side

drawer front

25mm (1")

6mm (¼")

1 Cut the sides and ends for the drawer boxes from strips of plywood of the correct height for each drawer. The drawer front overlaps the drawer box by 25mm (1") at the top and 6mm (¼") at the bottom. Glue and pin the boxes together: if you have a router, fit the drawer bottom into a groove for extra strength. Otherwise, glue and pin to the sides making sure they are square.

Assembling the drawer units

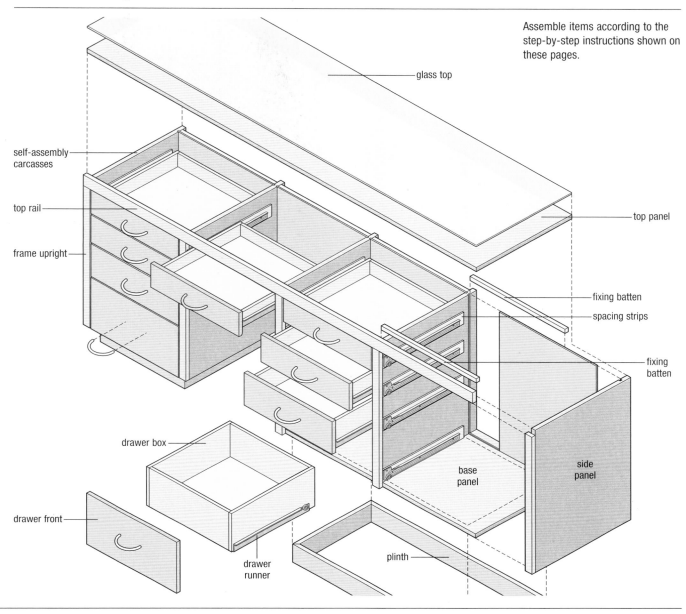

Assemble items according to the step-by-step instructions shown on these pages.

glass top

self-assembly carcasses

top rail

frame upright

top panel

fixing batten

spacing strips

fixing batten

drawer box

base panel

side panel

drawer front

drawer runner

plinth

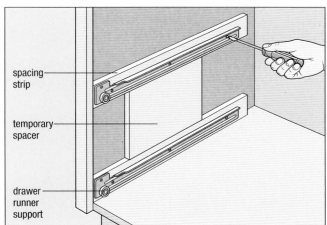

spacing strip

temporary spacer

drawer runner support

2 Fit the runner supports to the spacing strips on the side panels, using a temporary spacer to make sure they are level. Fit the corresponding runners to the drawer boxes and check they run smoothly. Make adjustments for correct alignment, then fix the drawer fronts neatly to the boxes, screwing through from the inside, and add the handles.

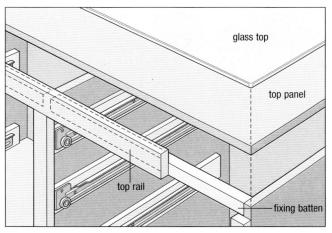

glass top

top panel

top rail

fixing batten

3 With all the carcasses in place, adjust their height if necessary to make sure the top edges are level. This is vital if fitting a glass top. Cut and fit the top panel, using fixing battens or **joint blocks** attached to the base units: set it down from the top edge of the top rail by the thickness of the glass. After painting the unit, lay the glass carefully in position.

Pine linen box

The design of this solid pine box makes good use of space in two ways. A storage chest for clean bedlinen is a useful addition to any bedroom, but whereas most blanket boxes open by lifting the lid, this design has a front-opening panel, termed a fall-flap. This allows for the unrestricted use of the top as a bedside table, or as an extra surface if the chest is placed at the foot of the bed, as in our picture.

You can make a box like this using laminated pineboard, a relatively new material, which is well suited for the home woodworker. Strips of seasoned pine are factory-glued in boards up to 600mm (24") wide, reducing the tendency to warp or shrink displayed by solid planks. They are machined and sanded ready for use, and can be stained, varnished or painted just like natural wood.

The overall size of the linen box can be tailored to suit your bedroom, but bear in mind that pineboards are supplied in standard lengths and widths, and if you design the box around these dimensions you can avoid unnecessary cutting, as suggested below.

This elegant, front-opening linen box also provdes an extra flat surface.

To enhance the finish of the storage box, avoid visible screw heads by using glued **dowel joints** to connect the main panels. The end grain of each board is concealed by adding a decorative moulding, available in a variety of carved or embossed patterns from most DIY stores.

MEASURING UP

The length of the box could be 900mm (36") to suit a single bed, or 1200mm (48") for a double.

- A suitable front-to-back dimension would be 500mm (20"). Pineboard, of this standard width, makes the sides, top and base without further cutting
- 600mm (24") is a convenient height from the floor, and the box can double as a seat
- Select a carved moulding to surround the top edges, and a less elaborate type such as **quadrant moulding** for the base. Match the height of the moulding to suit the board thickness (19mm/¾") in order to trim the end grain neatly
- The recommended hardware is obtainable from most DIY suppliers or mail order catalogues. Brass stays and hinges make a good contrast to the natural wood finish

MATERIALS

- Laminated pineboard 19mm (¾"): top, base, sides, back, front, shelf
- Softwood 50x25mm (2x1"): front battens 25x25mm (1x1"): shelf battens
- Moulding 19mm (¾"): to trim top and base
- Dowels 6mm (¼")
- Brass piano hinge 19mm (¾"): front panel
- Fall-flap door stays
- Auto-latches

TOOLS

- Basic Toolkit
- Dowel drill bit 6mm (¼") diameter with depth stop ●
- Hacksaw for piano hinge ●
- Mitre saw for mouldings ●
- Pin hammer & nail punch
- Sash cramps ●

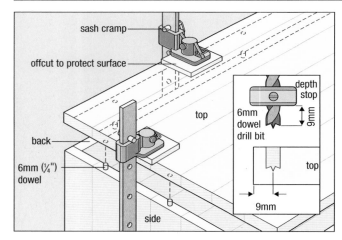

1 Cut the main panels from standard width boards. Mark out the dowel positions carefully: drill holes using a dowel drill bit with a depth stop *(see inset)* then transfer marks to adjacent panels with **dowel centre points** *(see page 29, step 2)*. Fix the shelf battens to the sides, insert the back panel, then add the top and base, gluing all surfaces generously. Use sash cramps to draw the joints together.

Assembling the pine linen box

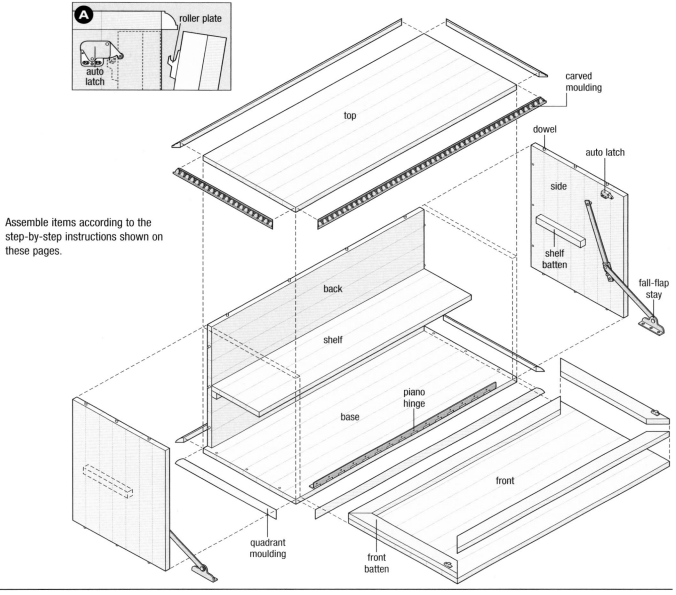

Assemble items according to the step-by-step instructions shown on these pages.

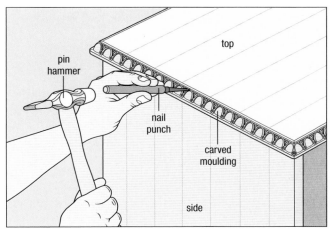

2 Wipe off excess glue immediately and leave to dry as long as possible (preferably overnight) for strong construction. Cut the mouldings to length using a mitre saw, or tenon saw and mitre box, and apply all round the edges with glue and panel pins. When glue is dry, round off the corners, and set the pin heads below the surface with a nail punch.

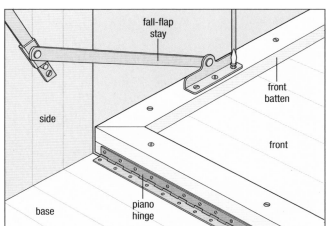

3 Brace the front panel with softwood battens, glued and screwed to the inside as shown. Screw the piano hinge along the bottom edge, offer the panel in place and fix with a few screws only. Adjust for a snug fit, with 1.5mm (1/16") clearance all round. Screw the hinge to the base, then fit stays and auto latches, which allow opening and closing without the need for a handle (see inset A).

Instant storage ideas for bedrooms

Adequate storage space in a small bedroom can be a large problem, so multi-purpose units, racks, stackable containers and anything that folds away when not in use are all invaluable. Underbed chests and bags are useful for storing items that are not in constant use, from extra blankets to ski-wear. Organizers fitted into drawers or hung in wardrobes are good for keeping clutter under control.

A clever rotating cupboard with one fixed and three adjustable shelves on one side and a full-length mirror on the other, also provides hanging poles that extend each side. With a square turntable of 406mm (16"), it will fit almost any bedroom.

A hanging belt hook saves drawer space, prevents belts from becoming entangled and makes it easy to find the one required.

A hanging tie rack with space for ten ties will prevent them from creasing and getting shut in drawers. It takes up very little space, and helps by displaying the tie fabric when matching it to a shirt.

A versatile, integrated system of different-size wire baskets, shelves and hanging rails can be used on its own or fitted into wardrobes. It is easy to adapt to any space and can be extended at any time as storage needs change. Once the top track is in position, components are simply clipped into position.

A handy drawer organizer which keeps control of items that are small, easily mislaid, tend to unroll or end up knotted round each other. The honeycombed dividers clip together and are ideal for drawers containing socks, tights, ties, belts or underwear.

Hanging containers can keep shoes and clothes tidy and accessible. From left to right: 24-pocket shoe organizer, 10-pocket shoe organizer, hanging wardrobe and 6-pocket sweater organizer.

This detachable laundry sack can be lifted off its frame and carried straight to the washing machine. It can be folded away when not in use.

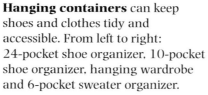

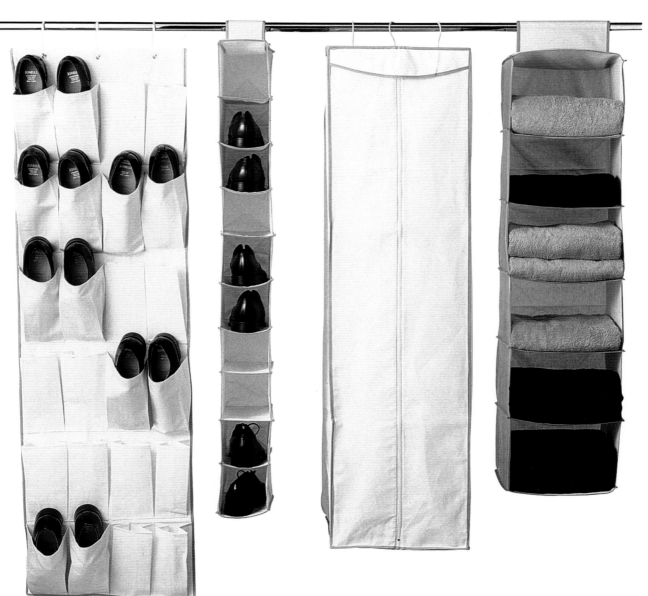

INSTANT STORAGE IDEAS FOR BEDROOMS

THE SPACE SAVER BOOK

CHAPTER 4

Bathrooms

An accumulation of bath oils, shampoo, soaps, shaving foam and so on quickly clutters up the bathroom, which is often one of the smaller rooms in the house. Even in a larger bathroom, storage can be a problem because the position of the plumbing restricts the scope. However, there is usually at least one wall where one or more narrow shelves, quite adequate for a mass of bottles and small containers, can be built. A corner shelf unit can provide similar storage space. Bulky items, such as towels, could be kept out of the way in baskets or even colourful painted buckets screwed to the wall through their bases. Where space to move around is limited, cupboards with folding or sliding doors can make access to their contents easier. Roller blinds would be equally effective space-saving 'doors' as well as attractive features. The wasted space around the basin can be put to good use by building a cupboard underneath, which will also conceal the pipework. For real style, particularly in a larger room, a complete unit the length of the wall, with a vanity top with the basin set in and cupboards and shelves underneath, will conceal bulky and unsightly items, while providing somewhere to display attractive containers and even houseplants.

PROJECT 1: *right.*
A stylish vanity unit with one double and two single cupboards beneath stores all bathroom paraphernalia tidily and safely. The curved open shelves at one end are also an attractive feature.

PROJECT 2: *below left.*
Shallow box shelving on an otherwise wasted wall behind the bath provides a convenient place to store bottles and containers and makes them into an appealing display. The shelves are tiled and the wood is varnished to waterproof the unit.

PROJECT 3: *below.*
A decorative cornice and glazed doors with sheer curtains turn this practical pine cabinet into an attractive centrepiece.

Fitted vanity unit

The benefits of a purpose made vanity unit for the bathroom, with ample built-in storage space below, can be achieved quite simply in your own home without resorting to expensive fitted units. Our picture shows a completely modernised bathroom layout, replacing an old-fashioned pedestal basin and free-standing cupboards or shelves with an integrated unit that is both practical and stylish. All unsightly plumbing is concealed under a vanity top with inset washbasin and splashproof surround, and the ample space beneath solves all the problems of providing a hygienic and comprehensive storage system for all your bathroom needs.

The requirements of a well-equipped bathroom need special consideration: a well-designed unit must allow for separate storage of a whole range of different items in compact and waterproof compartments: clean linen or used laundry, personal toiletry items or potentially hazardous cleaning fluids. All surfaces should be smooth and free of any sharp edges for reasons of safety, as well as being resistant to humidity and splashing water. Timber should be well primed and painted, and the unit raised off the floor on a plinth to prevent water damage. All hardware and fittings must be corrosion-resistant, and carefully positioned or concealed.

A stylish run of units for bathroom items including a fitted washbasin.

The gentle curve on the end of the unit shown here is both safe and practical as well as pleasing to the eye. With basic woodworking techniques, and a few professional pointers to the method for forming the curved fascia and shelves, this project shows how a smart and simple construction is perfectly achievable by the home woodworker.

MEASURING UP

When planning your new bathroom bear in mind the existing run of water and waste services to avoid unnecessary plumbing, though obviously where pipes do need to be extended they can be concealed within the base unit.

- It's a good idea to decide on the type of basin you want and offer it in place to check the location of the fitted top and the allowance required for access to plumbing connections *(see step 2)*
- Check that the basin is supplied with an oval template to assist you in making the cutout in the top
- Similarly, purchase the ready-made doors beforehand, and use their dimensions to plot the position of each compartment and the dividers on the base of the unit before starting
- Before making the template for the shelves, experiment with a sheet of paper or thin card to make a smooth curve, then transfer the shape to a sheet of MDF

TOOLS

- Basic Toolkit
- Edging clamps or G-cramps ⊕
- Hinge boring bit ⊕
- Jigsaw
- Router & profiling cutter
- Sash cramps ⊕

MATERIALS

- MDF:
  19mm (¾"): base, back, sides, dividers, shelves
  12mm (½"): template
- Laminated top and edging strip of your choice
- Planed softwood:
  100x25mm (4x1"): former battens for plinth
  75x25mm (3x1"): top rail
  50x25mm (2x1"): former battens for top corner assembly
- Plywood 3mm (⅛"): top fascia and plinth fascia
- Doors: ready-made panelled doors of suitable size, 25mm (1") thickness
- Plastic joint blocks *(see inset A opposite)*
- Concealed hinges and mounting plates *(see inset B opposite)*
  A hinge boring bit is usually obtainable from the same supplier

Assembling the fitted vanity unit

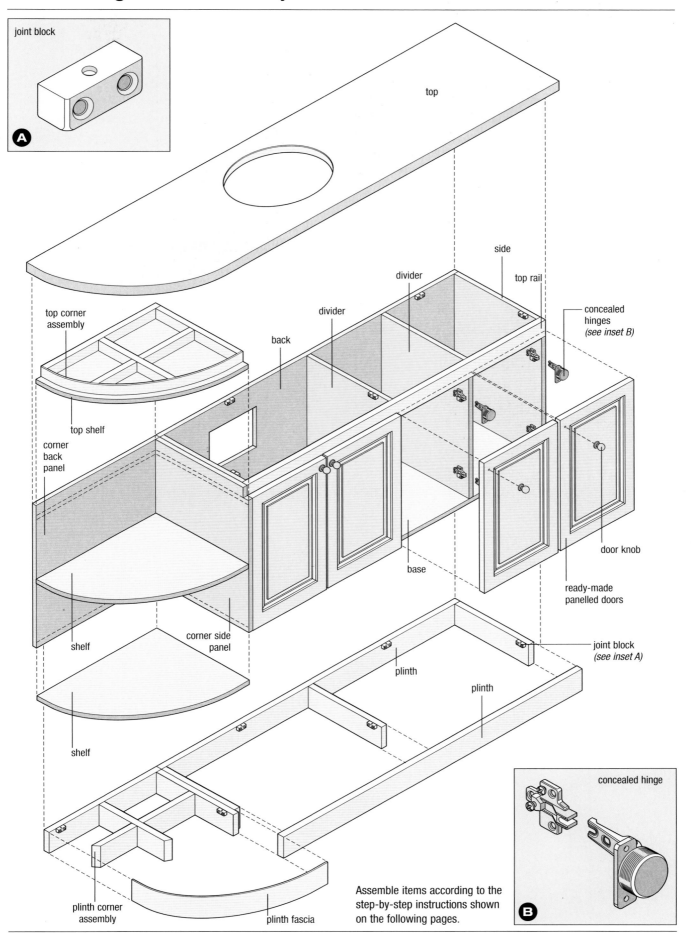

joint block

A

top

side

divider

top rail

concealed
hinges
(see inset B)

top corner
assembly

divider

back

top shelf

corner
back
panel

door knob

shelf

base

ready-made
panelled doors

corner side
panel

joint block
(see inset A)

shelf

plinth

plinth

plinth corner
assembly

plinth fascia

Assemble items according to the
step-by-step instructions shown
on the following pages.

concealed hinge

B

Constructing the fitted vanity unit

The method of assembly can be simplified by dividing it into several stages: the basic **carcass**, the corner unit, the plinth and the top. Most suppliers will pre-cut MDF boards to a standard width to suit the chosen size of your unit, which makes the basic carcass construction a straightforward operation. The corner assembly requires rather more care but is well worth the effort to produce a distinctive feature to round off the unit. Using a router safely is a skill worth acquiring for this project: when fitted with a **profile cutter** to follow a template it will produce a smooth curve and save a lot of work *(see step 4)*.

For the top, we don't imagine that your budget will run to solid marble as in our picture, so a laminated top with a waterproof surface is recommended: most

suppliers offer a wide range, in standard widths of 600mm (24") which should fit most washbasins. Check that the front edging strip you require is included with the worktop *(see step 10)*.

After making the cutout in the top for the basin, seal all the exposed edges of the chipboard core with waterproofing compound as recommended by the manufacturers to prevent moisture damage, and use a silicone-based mastic to seal the top in position. Similarly, thoroughly prime and undercoat all exposed surfaces and edges of the base unit to prevent water ingress, and finish with oil-based paint.

To complete the smooth lines of the unit the doors should be attached with concealed hinges which allow easy three-way adjustment.

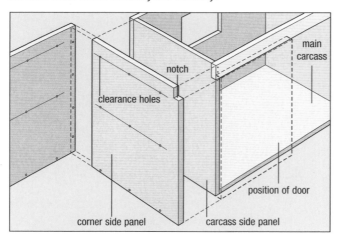

1 Begin the construction of the **carcass** by preparing the sides and dividers, which are identical except that the dividers are 19mm (¾") shorter at the bottom where they fit on the carcass base. Place all the panels together, flush at the top front corner, to mark out for the top rail *(see inset)*; then cut notches with a jigsaw. Always clamp the work to the bench for safety and accuracy.

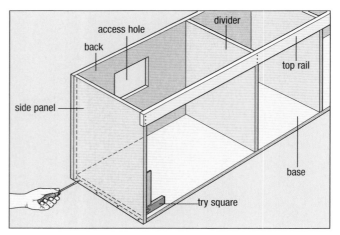

2 Before assembling the carcass, cut an access hole in the back to suit your plumbing connections *(see page 50, step 1 for tips on this procedure)*. Glue and screw the sides to the base, then add the back panel and the two dividers, positioned to suit the door sizes. Finally glue and pin the top rail in place, making sure all is square, and set aside for the glue to dry.

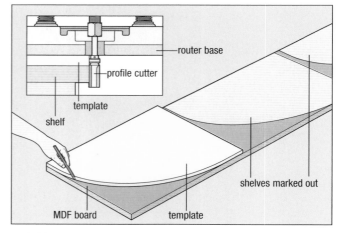

3 Prepare the two panels for the corner assembly: note that the corner side panel is wider than the carcass side panel by 25mm (1") at the front edge, to be flush with the doors. It also requires a similar notch at the top to finish flush with the top rail. Mark out and pre-drill **clearance holes** for the shelves as shown, but do not assemble the corner unit until the shelves are completed *(steps 4–5)*.

4 Make a template for the three curved shelves from 12mm MDF, using your paper pattern *(see Measuring Up)*. Lay the template on a board of 19mm (¾") material, draw round the curve and cut just outside the line with a jigsaw. Although you can smooth the curved edge by hand, by far the neatest way is to use a router with a **profile cutter**, which follows the curve of the template *(see inset)*.

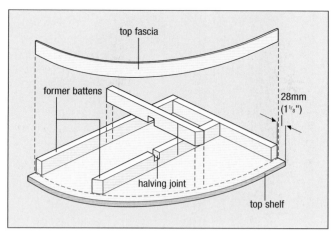

5 The top fascia is a thin strip of plywood or hardboard curved around **former battens** fixed to the top shelf. Draw a line 28mm (1⅛") back from the front edge, then glue and pin 50x25mm (2x1") battens, angled at the ends as shown, to support the bent plywood. Make a simple **halving joint** *(see page 37, step 6)* where the battens cross in the centre. Bend the fascia to shape as in step 7.

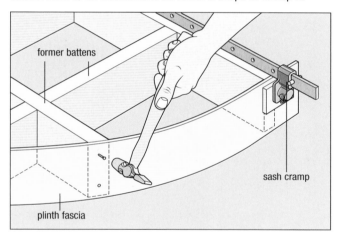

7 Construct the plinth, as shown on the main diagram, and make a former for the curved fascia as for the top shelf in step 5. Bend a strip of thin ply around the former battens and glue and pin in place. This is easier if you cut the strip longer than required and trim back when the glue is dry. Lay the main carcass on its back and attach the plinth using plastic **joint blocks** *(inset A on main diagram)*.

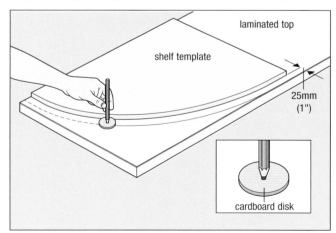

9 Cut the top to size, allowing a 25mm (1") overlap at the front edge. To plot the curve, use the template you have already made for the shelves: make a cardboard disc 50mm (2") in diameter, with a small hole in the centre for your pencil, and roll it around the template. This produces the larger radius required. Finally cut the hole for the vanity basin and seal round the edges as advised.

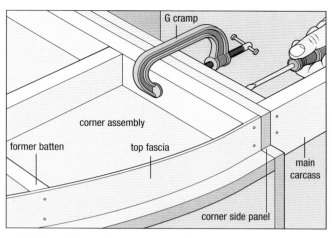

6 Glue and screw the corner assembly together, making sure it is square and all edges are flush. The front of the top fascia should align with the notch you have made in the corner side panel. Then attach the corner unit to the main carcass as shown. Apply filler to all joints and pinholes before fixing the top, and to all front edges. When dry, sand smooth ready for priming and painting.

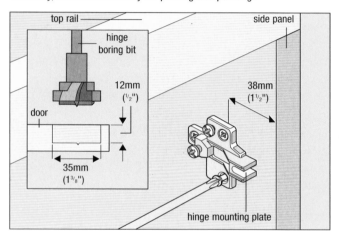

8 Move the unit into its final position and prepare the doors. The concealed hinges used here *(inset B on main diagram)* fit into holes bored in the doors with a special **hinge boring bit** *(see inset)*. They then locate on to mounting plates screwed to the carcass sides as shown. Alternatively, look for types of ready-made door which are supplied pre-drilled with the correct hinge mounting holes.

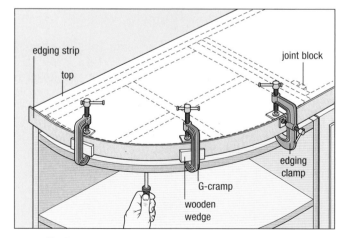

10 Fit the top to the carcass with plastic joint blocks, or use battens. Secure the curved end with long woodscrews up through the top shelf. Then fix the edging strip to the front edge with the recommended adhesive. Use special **edging clamps** for this, or improvise with small wooden wedges. When the glue is dry, trim the edging strip flush to the top with a sharp plane or file.

Bath storage unit

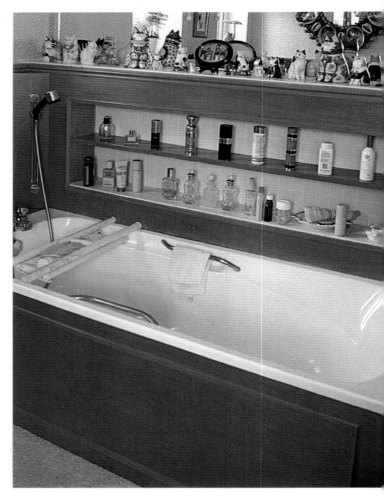

It is not uncommon when refitting or refurbishing a bathroom that services will be exposed to view. The solution is usually to fit cladding to the walls to conceal unsightly pipework, but this will inevitably take up valuable space. Here the otherwise wasted area between the bath and the rear wall has been put to good use by building in a shallow shelf unit, in the most convenient place within arm's reach of the bath.

This is a neat and attractive way of using every inch of space in a small room, and is enhanced by using veneered plywood, stained and varnished to match the bath panel. The interior of the shelf unit is tiled to seal against water which will inevitably collect there, and the whole construction finished off with an upper shelf along the whole length of the room.

This is a simple project, best tackled when replacing the bath, but the idea can be adapted to suit any position where panelling is fitted – to cover a concealed cistern, or when installing a shower, for example. Do ensure the components are thoroughly waterproofed both before and after construction.

Convenient, wipe-clean shelving keeps toiletries within arm's reach.

MEASURING UP

The depth of the shelf unit is determined by the space available, but also bear in mind the size of the tiles you will be using, and base the overall dimensions around this to avoid unnecessary cutting of tiles.
- Remember to add the internal width W of the moulding *(see inset A opposite)* to the size of the tiles when calculating the depth of the shelf
- If possible use a moulding the same thickness as the tiles for a flush finish at the front edge
- Depending on the overall length of the unit you may need to provide a central brace, to support the weight of the tiles

MATERIALS

- Veneered plywood:
  19mm (¾"): front panel, sides, top shelf, inner shelf, braces
  12mm (½"): sides, top and bottom of box assembly
  6mm (¼"): back of box assembly
- Angled hardwood moulding 22x22mm (⅞x⅞"): to trim
round box assembly and top shelf *(see inset opposite)*
- Angle brackets: to secure unit to wall
- Tiles, tile adhesive and waterproof grout
- Masking tape

TOOLS

- Chisel, craft knife
- Drill and drill bits
- Jigsaw
- Mitre box for mouldings
- Pin hammer, nail punch
- Screwdriver
- Tenon saw
- Tiling and grouting tools

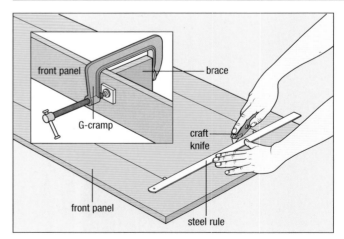

1 First cut a hole in the front panel for the storage box. Drill a 10mm (⅜") hole at each corner to take a jigsaw blade and score the grain of the plywood as shown to prevent splintering. Cut out the hole and square the corners with a chisel. Glue the **braces** in place *(see inset)*, add the sides and top shelf, and fix the whole assembly to the wall with battens or angle brackets as required.

Assembling the bath storage unit

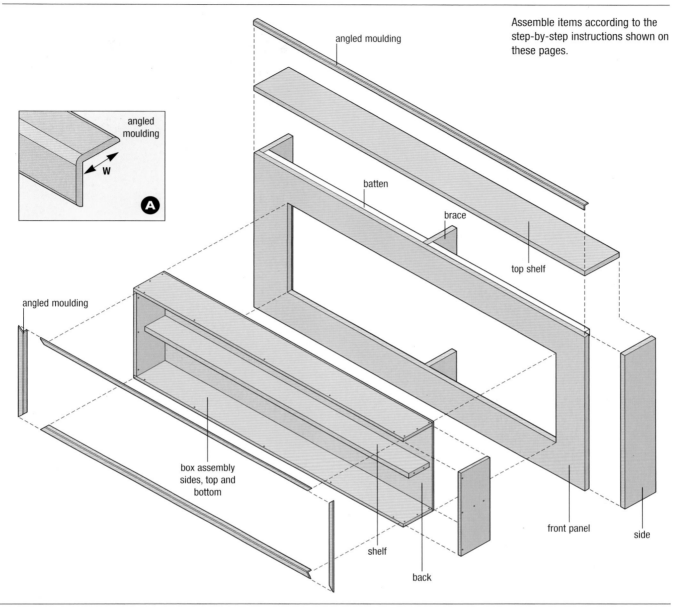

Assemble items according to the step-by-step instructions shown on these pages.

angled moulding

angled moulding

W

A

angled moulding

batten

brace

top shelf

box assembly sides, top and bottom

shelf

back

front panel

side

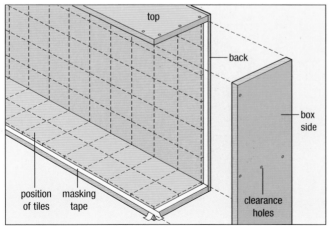

top

back

box side

position of tiles

masking tape

clearance holes

2 If laying tiles inside the box, it's easier to do this before assembly. Screw the back to the bottom, and apply masking tape around the edges to assist cleaning up excess adhesive. Allow to set, and meanwhile apply a coat of varnish to the other components, again masking the areas where they will be glued. Glue and screw the box together when dry, and drill **clearance holes** around the edges.

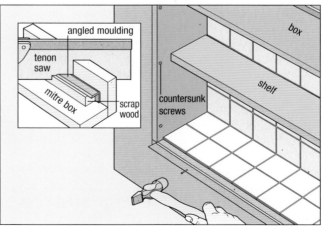

angled moulding

tenon saw

mitre box

scrap wood

box

shelf

countersunk screws

3 Screw the box into the front panel around the front edge as shown: the screws will be concealed by the angled moulding. **Mitre** the moulding and glue and pin in place. ★ When cutting mitres, use a block of wood under the angle of the moulding as shown *(see inset)*. This helps to keep it square and prevents splintering. Finally varnish all exposed woodwork to seal thoroughly.

Pine wall cabinet

The painted wall cabinet shown here has been cleverly designed to make maximum use of a small space and makes an attractive centrepiece to a well-designed bathroom layout. By making your own version you can include an extra touch of style, such as glazed doors or a moulded **cornice**, at minimal extra cost. Measure the height of the toiletries you use most often, and position the shelves accordingly, to get the most advantage of a custom-built design.

Made from easily obtainable materials, and carefully painted to seal the wood against moisture, this cabinet provides a practical and durable solution to any bathroom storage problem. Note how the two narrow doors provide good access without obstructing the washbasins, and are retained by magnetic catches for easy one-handed operation. By incorporating purpose-made wall hanging brackets into the design *(see inset opposite)*, it is possible to mount the cabinet safely on the wall without assistance, and adjust it afterwards to hang level so that the doors operate correctly.

Spacious but compact, this cabinet will hold toiletries for all the family.

MEASURING UP

Use planed softwood boards for the main cupboard to suit the depth: 150mm (6") square edged floorboards are ideal.

- Even though the recommended hanging brackets are adjustable, use a spirit level to set out the positions of the wall hooks before drilling holes
- Most timber suppliers stock standard **cornice** and **panel mouldings**: look for a suitable size to suit the proportions of your cabinet
- After you have constructed the doors subtract 3mm (⅛") clearance from the width and the height of the opening to give you the size of glass required

MATERIALS

- Planed softwood:
  150x25mm (6x1"): top, base, sides
  125x25mm (5x1"): top rail
  100x25mm (4x1"): shelves
  50x25mm (2x1"): door stiles and rails, frame uprights
- Plywood 6mm (¼"): back panel
- Dowels 6mm (¼"): for doors
- Softwood battens 22x22mm

- (⅞ x ⅞"): glass beading
- Panel moulding 38x16mm (1½x⅝"): for doors
- Cornice moulding for top
- Clear glass 3mm (⅛")
- Glazing tape
- Wall hanging brackets *(see inset A opposite)*
- Angle brackets for cornice
- 50mm (2") flush hinges
- Magnetic catches for doors

TOOLS

- Basic Toolkit
- Dowel drill bits ●
- Jigsaw
- Mitre box
- Spirit level

OPTIONAL:
- Router

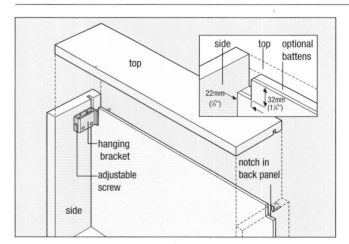

1 Before assembling the cupboard **carcass**, cut two notches in the back panel as shown, through which the hanging brackets will locate on the wall hooks. Rout 6mm (¼") grooves in the sides, top and base, 12mm (½") from the back edge, to locate the back panel. If you don't have a router, pin small battens around the back edge *(see inset)* and reduce the size of the back panel accordingly.

Assembling the pine wall cabinet

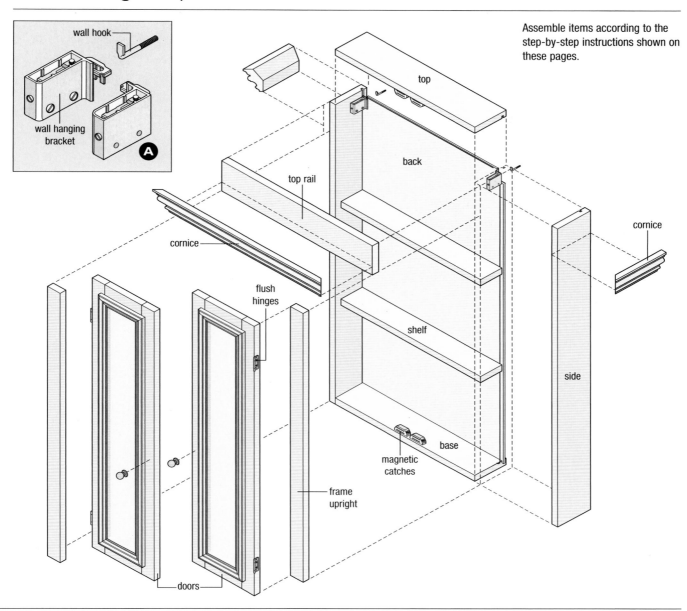

wall hook

wall hanging bracket

A

Assemble items according to the step-by-step instructions shown on these pages.

top

back

top rail

cornice

flush hinges

cornice

shelf

side

magnetic catches

base

frame upright

doors

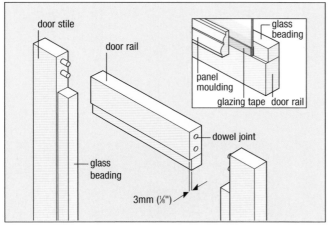

door stile

door rail

glass beading

panel moulding

glazing tape   door rail

dowel joint

glass beading

3mm (⅛")

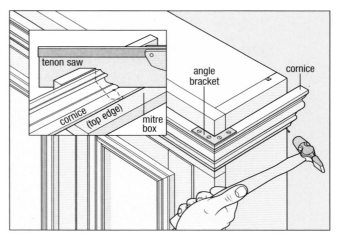

tenon saw

cornice

cornice (top edge)

mitre box

angle bracket

cornice

2 Glue and screw the **carcass** together and fix the top rail and frame uprights to the front. Make the doors using glued **dowel joints** *(see page 40, step 1)*. Glue and pin beading around the inside, leaving a 3mm (⅛") gap, fit the glass with **panel moulding** and **glazing tape** *(see inset)*, then fit the hinges and catches. Refer to page 37, steps 8 to 10 for further tips on glazing and hanging doors.

3 Fix the **cornice** around the top of the cabinet: glue and pin the bottom edges, and reinforce the top corners with small angle brackets as shown. The cornice should be cut at an angle in a mitre box to achieve the correct **mitre joint** *(see inset)*. Fix the wall hooks firmly to the wall, screw the hanging brackets to the cabinet, and lift into position. Use the integral adjustment screws to level it up.

53

Instant storage ideas for bathrooms

Space is often at a premium in bathrooms, so make the most of wall space with multi-level storage and with a variety of hooks. Any floor-standing units are more versatile if they are on wheels or castors, as they can be moved out of the way when the bathroom is in use or being cleaned. Look for waterproof materials. Chrome is traditional and plastics are colourful and easy to maintain.

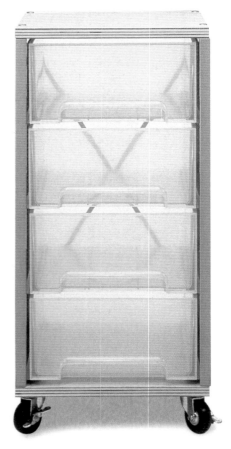

A plywood storage unit with deep polypropylene drawers is a neat solution to the problem of where to put bulky items. It is light enough to move easily on castors and its modern design would complement a contemporary decor.

A fold-flat clothes airer is useful on wet washing days and easy to store when it is not needed. It provides neat drying space for a full load of washing. A plastic-coated tubular steel frame ensures that it is sturdy and waterproof and it can be wheeled out of the way on castors.

Over-the-door hooks offer a quick and simple solution for storing towels and bathrobes. Made of steel and polymer, it locks together and rests over the door, so it is strong enough to support even the weight of soaking wet towels. The position of the hooks can be adjusted.

A corner tidy can be fitted into even the tiniest bathroom and can accommodate a surprising number of items on two shelves. Positioned near the basin or above the bath, it stores toiletries so that they are conveniently near to hand.

A neat chrome shower caddy has a place for everything necessary on its two shelves (one with a rail), soap dish and two hooks. It simply hooks over the showerhead itself and makes use of otherwise wasted space, providing not just storage, but also a helping hand.

This five-drawer, clear acrylic box is an ideal way of storing small, easily misplaced items, such as make-up, nail file and clippers, cotton wool and sachets of shampoo or bath oils. The contents remain clearly visible and so are easy to find.

A slim and stylish polystyrene and polypropylene cabinet with three shelves provides plenty of storage space. The rounded edges and sliding doors are useful safety features in a small bathroom. The toothbrush holder is convenient and hygienic and the double hook keeps towels within easy reach.

CHAPTER 5

The home office

Almost everyone needs a place to work at home, whether for a paid job, to pursue a hobby, to study or simply to write letters and file household documents. Some will be fortunate enough to have a room that can be set aside for the purpose, but with careful planning, a home office can be fitted into almost anywhere – from a teenager's bedroom to a corner of the living room. It can even be 'stored' in what amounts to a cupboard in an alcove, only coming into existence when folding doors are opened and a chair pulled up. A desk or worktable is fairly easy to construct and can be designed to fit both the available space and the particular requirements, from a sewing table to a desk for a designer.

Low-level filing cabinets or sturdy cupboards make excellent supports for a worktop, as well as providing storage. Shelves, either purpose-built or made from inexpensive kits with vertical metal tracking, are more or less a prerequisite and can be placed at heights suitable for the most commonly used items, whether reference books, ring binders or computer disks. In a small space or when the home office is part of another room, a desk and shelves can be combined into what looks like a single unit, with the storage space on the wall above the desk. Work- or hobby-related items can be quickly moved out of sight if you use a trolley or cabinet on castors.

PROJECT 1: *right.*
A shallow shelf under the desktops of this modular work station has been divided into compartments for different storage needs.

PROJECT 2: *below left.*
Two desktops, supported by freestanding shallow-drawer filing cabinets are positioned in an L shape. This combination provides a large working area, a smaller desk surface for the telephone and other equipment and unobtrusive storage space below.

PROJECT 3: *below.*
A plinth with open shelves supports the desktop in this student's bedroom, providing temporary storage for different study materials.

Modular workstation

An increasing number of people now choose to work from home for some or all of the time: most of them would agree that a specific area for their work is essential. Even if you are unable to devote a separate room to function as a home office, it should be possible to identify one part of your living space which can be set aside for the purpose. Whether you are running a full time business or just dealing with occasional paperwork, a work-station designed around the task keeps things in order and aids concentration by separating the work area from the surrounding distractions of domestic life.

Our picture shows the home office installed in a loft conversion, clearly with lots of room to spare. Any amount of space, however, will be of little use if it is not used efficiently. Designing your own workstation allows you to provide exactly the right mix of desk space and storage to suit your own needs. A **modular** system, using similar sized units designed around separate operations, allows maximum flexibility. It can be extended or re-arranged as required, and also looks smart and efficient: it can be surprising how much difference this makes to creating a productive and well-ordered office.

Building your own work-station also enables you to pay special attention to the ergonomics of the

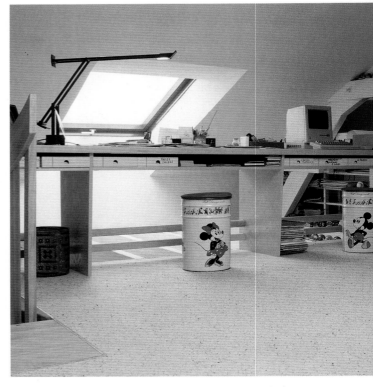

This modular design can be built, unit by unit, as the need arises.

design, often neglected in ready-made office furniture. A good posture is essential to health, especially for long periods of uninterrupted work. In considering this design, take into account the desktop height, keyboard height and reading distance (if using a computer), legroom and seating, and all the other variables which are unique to each person. Making your own purpose-built system puts these under your control and allows you to create the ideal working environment.

MEASURING UP

First decide on the profile for the sides which determine the height of the desktop. Note how the front edges of the side panels are cut away at an angle to enhance the visual impact of this design.

- A good working height is between 710-760mm (28-30"), always depending on the seating you will be using. When operating a keyboard your elbows should form an angle of 90 degrees for maximum comfort
- The width of each unit in a **modular** design is completely variable – those in the diagram are based around the width of a sheet of plywood (1220mm/48") to minimise the cutting required
- There is no reason why one or two of the modules could not be reduced in width to suit an item of office equipment, or to fit the remaining space at the end of the run of units
- Measure the most commonly used items in your filing system and incorporate purpose-made shelving or racking, in the most accessible position, in your design

TOOLS

- Basic Toolkit
- Bench plane ✚
- Block plane
- Drill and drill bits
- G-cramps
- Jigsaw or panel saw
- Pin hammer
- Router
- Screwdriver

MATERIALS

- Veneered plywood or MDF: 19mm (¾"): top, back, shelf 12mm (½"): sides
- Planed softwood 75x25mm (3x1"): rails
- Planed hardwood or softwood: 50x25mm (2x1"): front and back edging 25x6mm (1x¼"): cover strips, front trim 19x6mm (¾x¼"): shelf trim
- Cross dowels and bolts: for connecting rails
- 6mm (¼") dowels: locating dowels

★ Aligning cross dowels

After inserting the cross dowel in its hole *(see page 60, step 4)*, use a screwdriver to align the slot in the top with the bolt hole.

Assembling the modular workstation

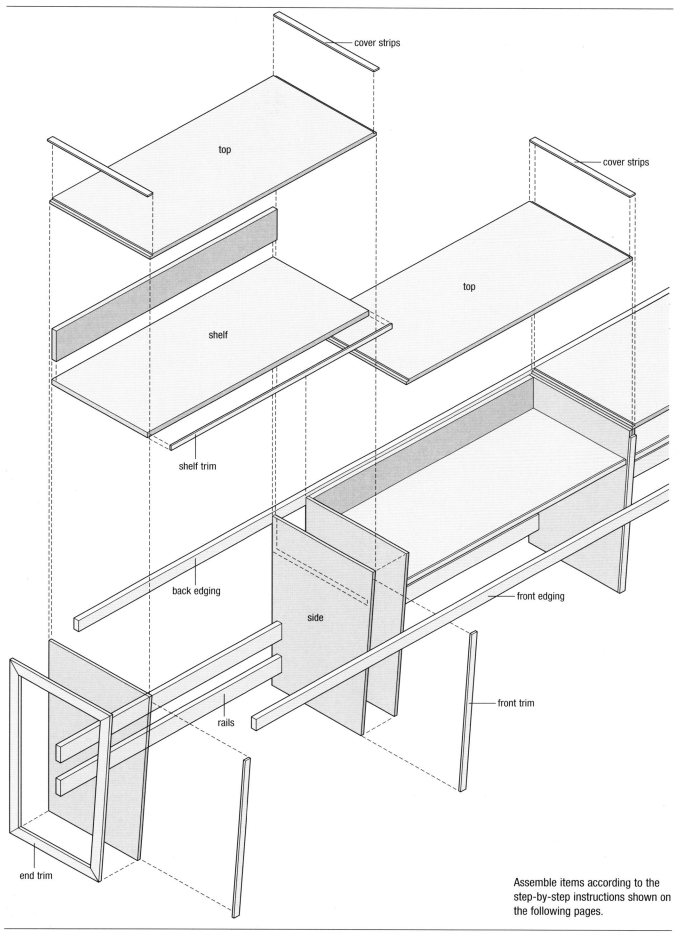

cover strips

top

cover strips

top

shelf

shelf trim

back edging

side

front edging

front trim

rails

end trim

Assemble items according to the
step-by-step instructions shown on
the following pages.

Constructing the modular workstation

The units illustrated here have been constructed using veneered plywood, varnished or lacquered for a really polished finish. Although the material is more expensive than plain MDF, the resulting finish is really worth it, and still achievable at half the cost of buying ready made office furniture. The solid wood edging used to connect the units and trim the shelves should ideally be of the same type of hardwood to match the veneer – usually available ready planed from the same supplier. If you don't want this option, planed softwood, painted or varnished, will be quite adequate.

Decide on the number of modules required, and prepare all the side panels at the same time. Each module is made to an identical profile, and the sides, although only 12mm (½") thick, form a strong support when fixed together in pairs, the edges being connected with the front trim. This should be pinned in place without using glue, to allow the units to be separated. The entire work area can thus be moved around or extended at some future time.

The strength of each module also depends on the **cross dowels** which connect the rails to the sides *(see step 4)*. These threaded brass inserts, obtainable from hardware stores or mail-order suppliers, make it possible to bolt the units firmly together without using glue, and take them apart later if required. The simplicity of this design uses the minimum of components and is completely adaptable to the needs of any user, whatever the space available.

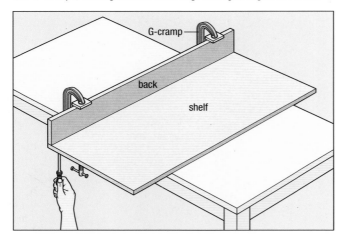

1 Decide on a comfortable working height for the desktop, as explained on the previous page, and cut the sides to shape. Draw an outline for the side profile as shown, and remove waste with a jigsaw. Clamp the sides together in pairs to plane the front edges for a perfect match. Use a workbench vice if you have one, or improvise with two G-cramps, making sure all three edges at right-angles are flush.

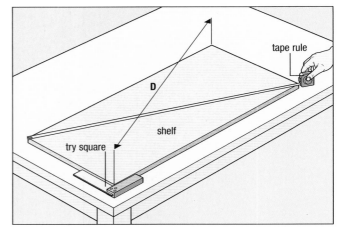

2 Cut the shelves to size, making sure they are absolutely square before proceeding. Care at this stage makes all the difference to an accurate assembly when two or more units are joined together. Even if using panels pre-cut to size by the supplier, check all round with a try square, and take a measurement 'D' from corner to corner – when the diagonals are equal the corners should be square.

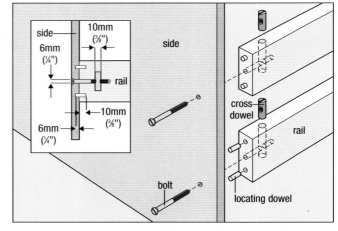

3 Glue and screw the shelf to the back panel as shown, again making sure they form a right angle for accurate assembly. Drill **pilot holes** for the screws, countersink the heads, and set aside for the glue to dry before proceeding. Make the shelf deep enough to accommodate your files or other office equipment, but take care not to reduce the clearance underneath for adequate leg-room.

4 The rails which link the sides also act as a foot-rest, so must be securely fixed. Use countersunk bolts and **cross dowels** in the ends of the rails *(see Materials)*. Drill a 10mm (⅜") hole in the top of each rail to receive the threaded dowel, then matching 6mm (¼") holes in the ends of the rails and the side panels. Glue locating dowels in the rails to engage in holes in the sides and prevent twisting.

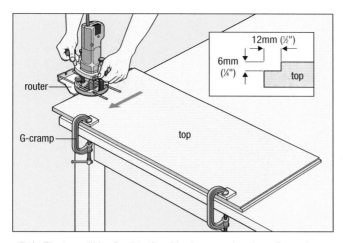

5 The top will be fixed to the sides by screwing down from above, but if you have a router you can refine the procedure to produce a flush surface with no obstructions, ideal for use as a desktop. Use a router fitted with a side fence to **rebate** both ends of the top panel, 6mm (¼") deep and 12mm (½") wide *(see inset)*. This allows the fixing screws to be concealed by a cover strip after assembly.

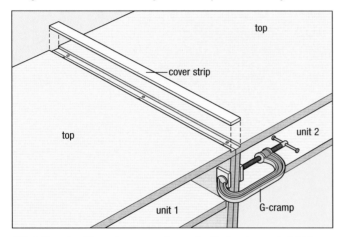

7 You should now have a neat run of units forming a straight line with a level worksurface. If the floor is uneven, insert small wedges under the side panels where necessary to bring all the tops level. Clamp each pair of units in turn, making sure the front edges are flush, and screw the sides together. Finally insert the cover strips to cover the join between each unit and pin into place.

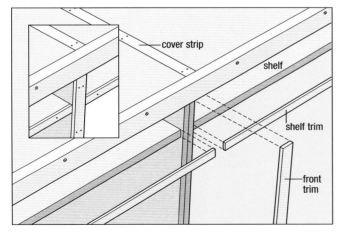

9 The end grain of the sides and shelves is concealed using 6mm strips of planed timber to match the finish of the units, whether veneered or painted. Cut shelf trim to length and pin in place. Do likewise for the front trim, which covers the join between each pair of side panels and locks them together. As with the cover strips in the desktop, omit glue in case the units ever need to be separated.

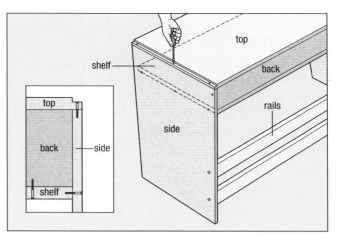

6 Assemble each unit in turn, fixing the top and the shelf as shown *(see inset)*. Cut a spacer, the same height as the back panel, to help locate the shelf parallel to the top. Fix the rails with countersunk bolts, as described in step 4, to draw the sides tight together for a really solid construction. When all the units are assembled, line them up in their final position ready to link together.

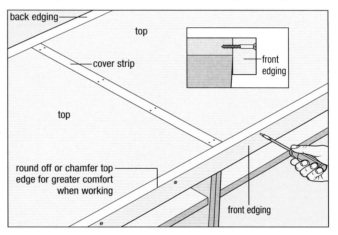

8 Use a block plane to adjust the front of the desktop to be perfectly flush, and fix the front and back edging with countersunk screws *(see inset)*. If at all possible, use battens long enough to cover the whole length of the finished unit in one continuous run. Otherwise stagger the lengths so that the fixing screws lock the units together each side of a join as illustrated.

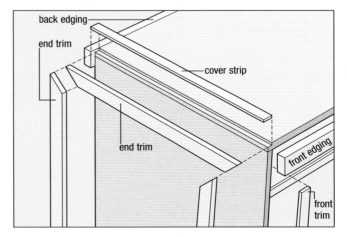

10 **OPTION:** At each end of the run of units, the cover strip and front trim can be reduced from 25mm (1") to 12mm (½") in width, as there is no adjoining side panel. However you could pin strips of 12mm (½") timber, mitred at the corners, in place of the next unit. This option maintains the visual effect of the modular system, and makes a neat finish to the whole assembly.

Portable desktop

The ultimate in space saving for the home office is a portable desktop, simply placed on top of two or more filing cabinets. When not required it can be removed and tucked against the wall until later. The example in our picture is a little more sophisticated – the arrangement of storage cabinets lends itself to an L-shaped worksurface, giving easy access at one side to the essentials of office equipment, whilst leaving a clear working area ahead.

The L-shaped desktop can still be dismantled and stored flat to save space, by using the **panel connectors** shown in the inset opposite. These useful items allow two panels to be drawn tightly together with a few turns of a small spanner. They are concealed from view, recessed into the underside of the desktop, and yet easily accessible when required.

An even quicker way to produce the instant desktop is to use a pair of doors laid flat on suitable supports. The less expensive lightweight doors with a hollow core are not suitable for fitting panel connectors, but heavy duty **door blanks** with flush

A versatile arrangement of worktops that stow away when not in use.

plywood surfaces and solid core are obtainable from most timber merchants. These may be already fitted with hardwood edging on the long sides, but the ends will probably need to be edge trimmed as in step 3.

MEASURING UP

This simple project requires very little measurement or preparation. If making your own desktop from plain or veneered MDF, use 25mm (1") thick material if possible for a really solid worksurface.

- Make the edge trim from thicker material (38-50mm/ 1½-2") using softwood or hardwood battens. When glued and pinned around the edges they stiffen the panels considerably and prevent bowing
- Ready made door blanks are available in sizes up to 2130x910mm (7x3'), perfect for a large solid desktop
- If supporting the desktop on a pair of filing cabinets, for example, you may need to adjust the height to achieve a comfortable working position. Screw blocks of wood to the underside of the desktop if necessary
- Alternatively, rest the desktop on a pair of folding trestles which can be obtained very cheaply, allowing the whole assembly to be stowed away when not required

MATERIALS

- MDF 25mm (1"): top, side
- Planed softwood: 38x19mm (1½x¾"): edge trim
- Panel connectors
- 10mm (⅜") dowels

TOOLS

- Block plane
- 6mm (¼") chisel
- Drill and drill bits
- Hinge boring bit ⊕
- Mitre box
- Pin hammer
- 10mm (⅜") spanner
- Tenon saw

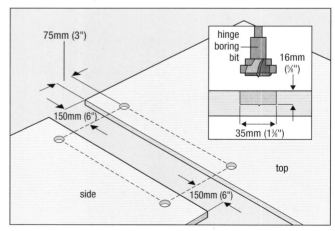

75mm (3")

150mm (6")

hinge boring bit

16mm (⅝")

35mm (1⅜")

top

side

150mm (6")

1 Take special care to plane the edges straight and square where the top and side panels meet. When satisfied that you have a neat joint, turn the panels over and mark out the positions of the **panel connectors** to the dimensions given above. Bore four holes 16mm (⅝") deep, using the same special **hinge boring bit** as used for mounting concealed hinges (*see Fitted vanity unit, page 49, step 8*).

Assembling the portable desktop

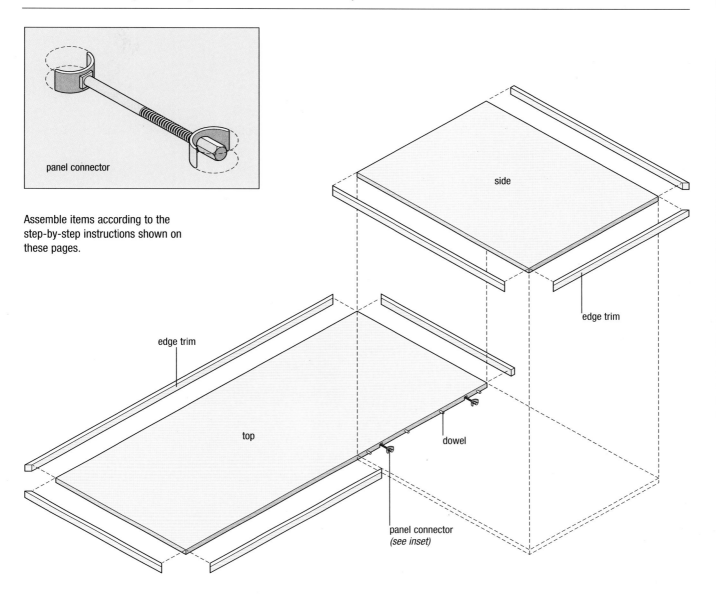

panel connector

Assemble items according to the step-by-step instructions shown on these pages.

side

edge trim

edge trim

top

dowel

panel connector
(see inset)

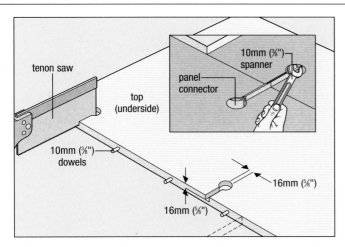

tenon saw

panel
connector

10mm (⅜")
spanner

top
(underside)

top

10mm (⅜")
dowels

16mm (⅝")

16mm (⅝")

16mm (⅝")

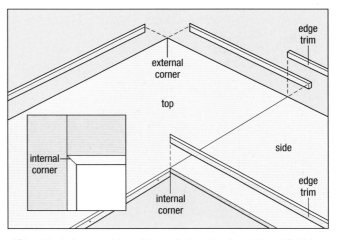

edge
trim

external
corner

top

internal
corner

side

internal
corner

edge
trim

2 Make two cuts with a tenon saw at each hole, and chisel out the waste to form a channel for the connecting bolt. To ensure that the top surface is flush, use **dowels** to align the two panels – follow steps 2 and 3 on page 29 for making **dowel joints**. Turn the panels over and place in position; insert the connectors and tighten the bolts to draw the panels together for a really strong joint.

3 Attach the edge trim with panel pins, glued and pinned flush with the top edges. **Mitre** the corners, beginning at the internal corner where the two panels intersect *(see inset)*. Shave the mitre joint for an accurate fit, then mark out and cut the external mitres, working around the desktop. Note how the trim is cut to fit the two panels separately to allow for their disconnection when required.

Easy-build desk

If the large **modular** workstation described on pages 58-61 is not for you, this small desk, easily constructed from inexpensive materials with the most basic woodwork techniques, might be more suitable. It is light enough to move around the room single-handed, and can be purpose-made to suit any task. There are a number of occasional uses for which a specially designed working area comes into its own: domestic paperwork, home studies, crafts or hobbies, for example. Such activities have similar requirements: a well laid-out worksurface at a comfortable working height, and a dedicated storage space where books, stationery or tools and materials can be safely stowed until needed.

This project uses the same construction techniques and hardware as other self-assembly or flat-pack furntiure, with the added advantage that you can design the shape and size around your own requirements. The unit occupies the minimum floor area, with the worktop cut away at an angle to provide generous legroom and a comfortable seating

Perfect for a small work area: compact but useful; stylish but simple.

position. The tapered profile also allows easier access to the shelving on the right side of the workspace.

MEASURING UP

The shape of the angled top may be determined to suit your own purposes, bearing in mind the floorspace available and the layout of the rest of the room where space may be limited.

- Try out various seating positions to establish which angle gives the most comfortable legroom before cutting the profile on the desktop
- The overall depth of the desk, and the small column of shelves, has been designed around standard sized furniture board as available from all DIY stores
- Common widths are 300mm (12"), 375mm (15") and 450mm (18"), all of which can be combined to make a unit of similar proportions to this one
- All you have to do is cut the panels to length to make a convenient height – some suppliers will even do this for you at no extra charge, or very low cost

MATERIALS

- Furniture board 19mm (¾") or 15mm (⅝"): top, end, back, shelves, plinth, shelf column
- Connecting bushes (A): for assembling the unit
- Plastic joint blocks (B)

TOOLS

- Block plane
- Drill and drill bits
- Jigsaw
- Screwdriver
- Tape measure
- Try square

end — 32mm (1¼")

back

50mm (2")

16mm (⅝") **A**

back

6mm (¼")

back

100mm (4")

1 Prepare the back, end and column side by drilling holes and inserting four **connecting bushes** (A) as shown. These are a smaller version of the **cross dowels** used in the Modular workstation, page 60, step 4. Note how, while the right hand shelf column supports the desktop, the left hand end panel is slightly higher. Allow for this extra 32mm (1¼") when marking out for the connectors.

Assembling the easy-build desk

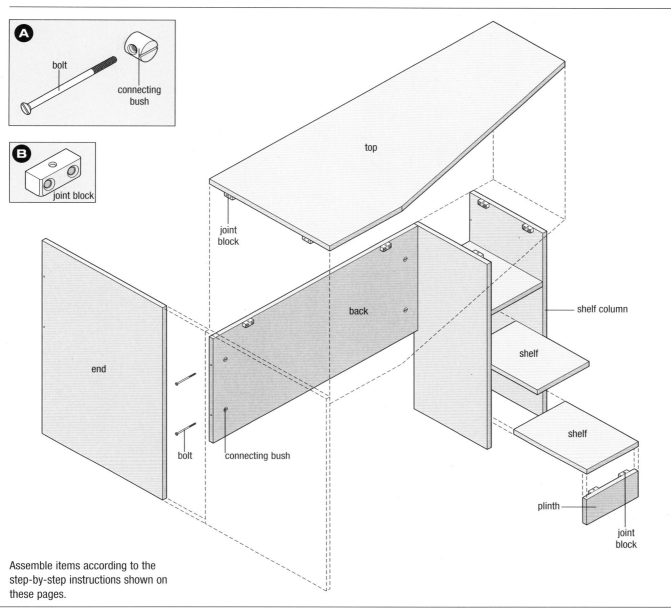

A
bolt
connecting bush

B
joint block

top

joint block

back

end

bolt

connecting bush

shelf column

shelf

shelf

plinth

joint block

Assemble items according to the step-by-step instructions shown on these pages.

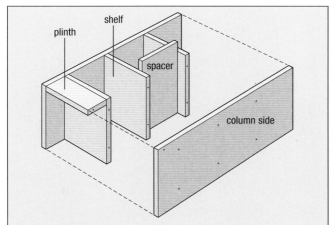

plinth

shelf

spacer

column side

2 Assemble the shelf column before finally connecting the side and end panels. Cut the plinth and shelves from a single strip of board to produce a unit of constant width. First join the plinth and bottom shelf with plastic **joint blocks** (see inset B above): the plinth raises the shelf off the floor and also keeps the unit square and rigid. Then fit the sides, using a spacer to position the shelves equally.

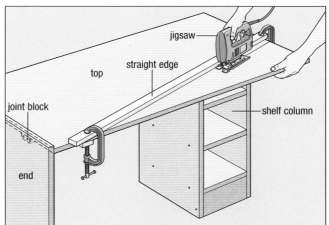

jigsaw

top

straight edge

joint block

end

shelf column

3 Assemble the desk with the connecting bolts, and fix the top with plastic joint blocks. Mark out the chosen angle on the desktop and remove the waste with a jigsaw, using a straight edge clamped to the top as a guide. Clean up the front edge with a block plane, and round off the corner for safety. Fill the edge grain with a proprietary woodfiller and paint to match the coated surface.

Instant storage ideas for the home office

Gone are the days when the efficient office was a uniform dreary grey. Now inexpensive, stylish and colourful containers can make filing almost a joy. Stackable cardboard or plastic boxes, trolleys with sliding baskets, filing cabinets, even waste bins come in bright colours. Attractive rattan trunks could also provide voluminous archive storage for material that is not currently required.

This black metal rack is ideal for storing newspapers, magazines and manuals that can easily be torn, even on bookshelves.

A magazine tidy made of sturdy cardboard can also store exercise books and files. The handles ensure that it is easy to lift when full.

A selection of filing and deed boxes. From top to bottom: brass and galvanized steel, opaque plastic, galvanized steel, cardboard drawers with woven fabric handles, cardboard with metal-reinforced corners and moulded polythene.

Cardboard box files in a wide range of heights, widths and depths can be used to store just about everything in the home office, from artwork to fabric swatches. Labels can be slotted into holders on the front, so that the contents can be identified at a glance.

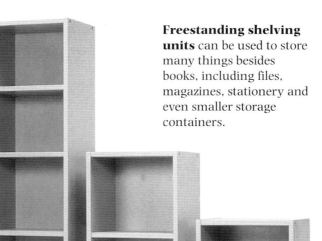

Freestanding shelving units can be used to store many things besides books, including files, magazines, stationery and even smaller storage containers.

A complete work station can be designed from individual modules to suit all home office needs: vertical track shelves above and below a desktop and basket drawers beneath it.

A sturdy, smooth-running metal trolley is useful if the home office is part of another room. Work- or hobby-related items, such as box files, computer equipment or a sewing machine, can easily be moved out of the way when work is over.

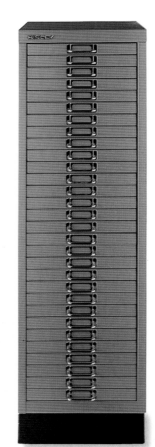

This red metal, multi-drawer filing cabinet is a convenient way of storing stationery, small pieces of equipment or easily damaged items, such as photographs.

CHAPTER 6

Hallways

A large hall could do double duty as a dining room, office or study. A small hall is often little more than a narrow passage with no room for floor standing furniture. Here shelving and hanging storage come into their own. The narrowest hall can take a single shelf, fixed around the walls at picture rail height. In corners, shelves or even hanging baskets will provide further storage. Alternatively, an arrangement of semi-circular garden containers or old-fashioned bicycle baskets can be fixed along a wall. A doorway at the end of a passage could be framed by a box shelving system. Coat stands look ornate but they take up floor space. A coat rack screwed to the wall is a slimmer, neater alternative.

The space under the stairs is often long and narrow, sometimes boxed-in, with a door at one end. This makes it disastrous for storage as once the floor is cluttered with miscellaneous belongings it becomes impossible to find anything. By removing the side partition and using the space from this angle a much more efficient storage area becomes available. If you are able to stand upright in part of it the space under the stairs may be suitable for use as a home office, shower, cloakroom or laundry area.

PROJECT 1: *right.*
Build a set of pull-out shelves for ample storage of household linen or clothing. Smooth gliding castors provide easy access.

PROJECT 2: *below left.*
Part of the understairs space is a storage cupboard, while the remaining area is fitted with tailor-made shelving to accommodate an extensive library of books.

PROJECT 3: *below.*
The minute and awkward space under these stairs provides an ideal additional storage area for an ironing board and other household belongings. The unique design of the cupboard door means that it is hinged twice so that it completes a right angle, enclosing the cupboard space.

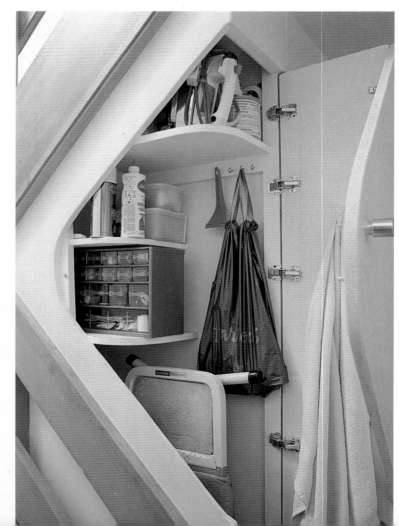

Understairs storage units

Most forms of understairs storage share a common problem: the more inaccessible the storage area, the more difficult it is to keep organised and presentable. Given that the hallway is the one area of your home that is likely to be seen by the greatest number of people, this is hardly the best state of affairs. The project we have selected is an ingenious and practical way of dealing with this problem, and is bound to attract favourable comment from the most casual visitor. At the same time the storage units when stowed away are quite unobtrusive and blend naturally with the staircase design. By adapting the finish of the front panels to complement your existing decor you can turn a useful storage idea into an attractive feature.

The shape of your staircase will determine how successfully you can utilise the space beneath it in this fashion. The one in our picture has a quarter-landing near the bottom which gives a suitable height for the smallest trolley. If your stairs are what is called a 'straight flight', running directly to the floor in a straight run, you can adapt this design by fitting a triangular filler panel at the bottom to a height of 250mm (10") *(see page 72, step 1)*. If your flight is shorter, omit trolleys to suit the arrangement. For practical purposes the width of

These efficient, pull-out units give access to an often awkward space.

each trolley should not exceed 600mm (24").

Depending on the layout of your hallway, you may decide to fit a partition at the higher end to conceal the side of the tallest trolley. This is simply done by constructing a **stud partition** from 50x50mm (2x2") timber fixed to the wall, floor and ceiling, and cladding it with sheet material such as plasterboard or wallboard.

MEASURING UP

Measure the usable space **L** *(see page 72, step 1)* and decide on the number of trolleys which will fit under the stairs.

- You should allow a clearance of 6mm (¼") between each trolley and at each end
- Calculate this allowance and subtract it from the overall length **L**
- Divide by the number of trolleys required to give the overall width of each front panel
- Try to achieve a standard width for each trolley of 500-600mm (20-24") for maximum stability. If this is not possible you may have to adjust the size of the filler panel at one end
- If fitting **hockey-stick moulding** or other edge trim *(see detail A opposite)*, remember to reduce the actual width of the front panels accordingly
- It's a good idea to proceed with assembly of the bases *(see page 72, steps 1-4)* before calculating the height of each unit *(page 73, step 5)*

TOOLS

- Basic Toolkit
- Adjustable bevel gauge ●
- G-cramps
- Jigsaw
- Mitre saw if available ●
- Plumb line & bob
- Round file or rasp ●
- Spirit level

MATERIALS

- MDF 19mm (¾"): front and back panel, bracing panel, bottom shelf, castor blocks
- Planed softwood:
  75x25mm (3x1"): slats, bracing batten
  50x25mm (2x1"): facing strip
  25x25mm (1x1"): mounting battens, slat support battens, mounting block
- Moulding 19x6mm (¾x¼"): rubbing strip

- Fixed-wheel rubber-tyred castors 75mm (3") diameter
- Rubber doorstops (2 per unit)
- *OPTIONAL:*
- Hockeystick moulding 19x6mm (¾x¼"): edge trim
- Polycarbonate sheet 3mm (⅛") thick: viewing panel
- Bullnose moulding 12x6mm (½x¼") and square moulding 6x6mm (¼x¼"): window panel beading

Assembling the understairs storage units

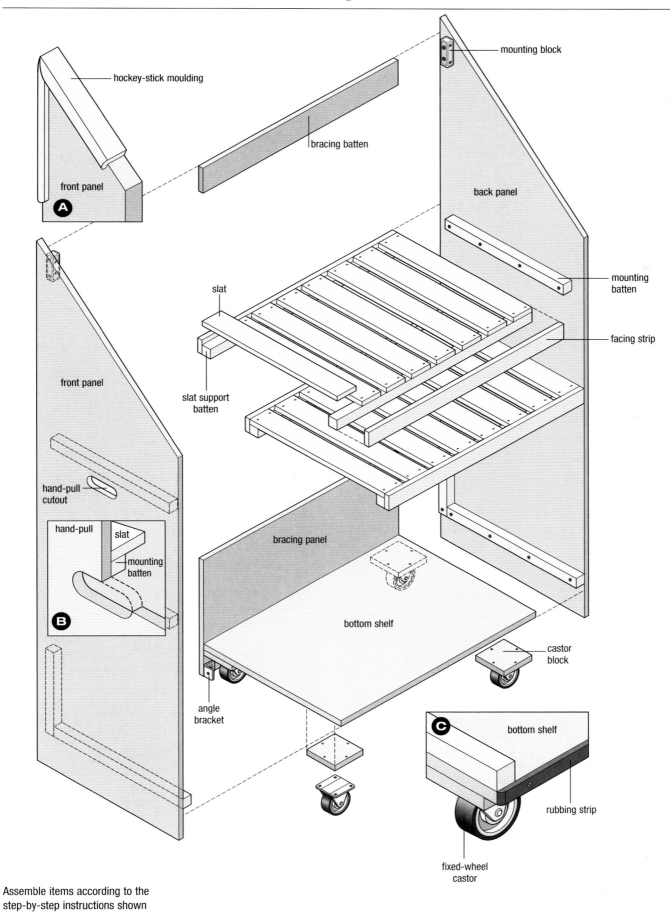

hockey-stick moulding

front panel

A

bracing batten

mounting block

back panel

mounting block

mounting
batten

slat

facing strip

front panel

slat support
batten

hand-pull
cutout

hand-pull   slat

mounting
batten

B

bracing panel

castor
block

bottom shelf

angle
bracket

C   bottom shelf

rubbing strip

fixed-wheel
castor

Assemble items according to the
step-by-step instructions shown
on the following pages.

Constructing the understairs storage units

When planning this project, a practical point to note is that the height beneath the stair string reduces by almost as much as the distance moved towards the bottom of the stairs: most domestic staircases are constructed to a standard 40° rise. The positioning of the trolleys is therefore quite critical. To maintain the correct spacing, pin a 'rubbing strip' to the bottom shelf of each unit: don't glue them on, in case you need to renew them at a later date *(see step 2)*. Using fixed-wheel castors will ensure that the trolleys run parallel and large, wide wheels will make the operation easier.

In most cases the flooring in your hallway will continue into the under-stair area, but if for some reason it does not, or if you have a very uneven floor, you would be advised to fit a floor panel to provide a smooth and level surface for the castors. Use a sheet of hardboard or plywood of a suitable thickness to bring the floor level up to match that in the main hallway. If the hallway is carpeted, make the floor panel roughly the thickness of the underlay and you should find the trolleys run quite smoothly over the join. If necessary, **chamfer** the front edge of the floor panel.

A good finishing touch is to fit a pair of rubber doorstops to the back of each unit to prevent them knocking against the rear wall. If you choose this option, remember to allow for the thickness of the doorstops before calculating the depth of the unit.

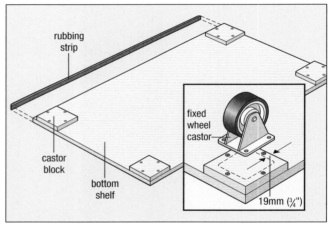

1 Drop a plumb line from the inner edge of the upper newel and mark point X on the floor. Measure distance L from here to the lower newel Y (or to the filler panel Z on a straight flight). Work out how many units will give a front panel width of 500 to 600mm (20 to 24"), allowing 6mm (¼") clearance between them and at each end. Finally, measure to the rear wall for the overall depth of the units.

2 Cut a bottom shelf to the overall width of each unit less the thickness of the bracing panel. Fit a castor block at each corner, inset on one side by the thickness of the **bracing** panel. Pin the rubbing strip along this side. (Not using glue allows for easy replacement). Screw the castors centrally to the blocks, exactly parallel to the edge of the bottom shelf.

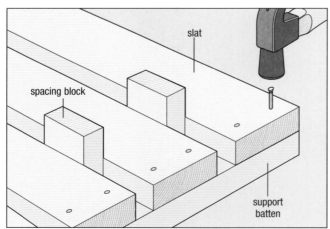

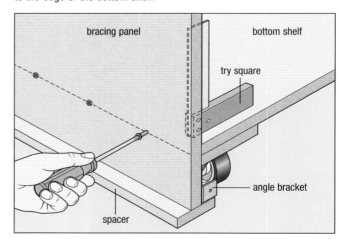

3 To make the shelves, cut the required number of slats all the same length from 75x25mm (3x1") softwood – round off the edges before fixing. Glue and pin the slats to the support battens, using offcuts of timber to determine even spaces between slats. Make sure the corners are square before fixing the facing strips: use the bottom shelf as a template.

4 Screw an angle bracket to the bottom corner of the bracing panel. Stand the panel on a 10mm (⅜") strip of timber as a spacer and mark the centre line of the bottom shelf. Drill a line of countersunk **clearance holes** to fix the panel to the edge of the shelf. Hold it square with a try square and glue and screw into position. Repeat this operation for all the base units.

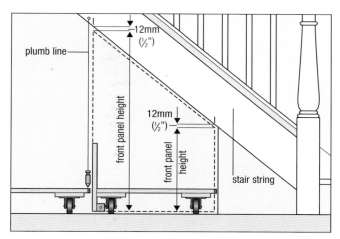

5 Slide the bases into position under the stairs to check they are a good fit. Using a plumb line, mark the stair string in a vertical line with each edge. Measure from the bottom of the bracing panels to the stair string, subtract 12mm (½") for clearance, then mark out and cut the front and back panels. Cut the top edges with care – they should all be the same angle.

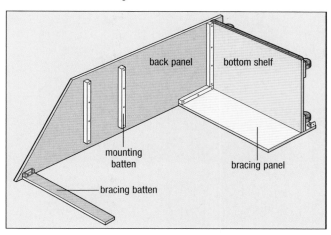

7 Assemble the unit. Place the base assembly on its side on a flat surface and slide the back panel to meet it, locating the mounting battens. Glue and screw together and do the same for the front panel. Then add the shelves in turn, working towards the top, and finish by screwing the two mounting blocks to the bracing batten. Make sure the assembly is square and let glue dry before raising into position.

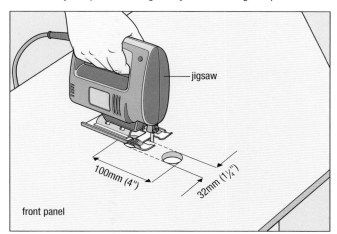

9 Make a cutout in each front panel for a hand-pull. Drill two 32mm (1¼") holes 100mm (4") apart and remove the waste with a jigsaw. Make sure the top edge of the cutout is flush with the mounting batten behind, as shown on the exploded diagram (inset B). Smooth the edges with a round file or rasp and finish off with fine sandpaper.

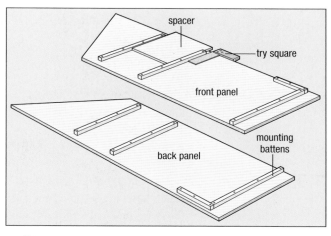

6 Having decided on the shelf positions, fix the mounting battens to the front and back panels beforehand for easy assembly. Glue and screw in place, using a spacer to locate the battens for the slatted shelves a constant distance apart: the layout of the back panel should be a mirror image of the front. Fix the mounting battens for the bottom shelf and bracing panel in the same way.

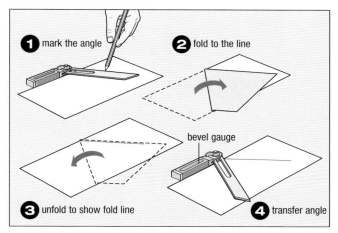

8 A **hockey-stick** moulding (see detail A on assembly diagram) provides a neat finish to the front panel. To make a neat **mitre joint** at the corners you will need to measure the angle, halve it, and use an adjustable bevel gauge to transfer it to the moulding.
★ To save calculation, a simple short cut is to use a piece of thin card as shown above. Cut the joints with care using a mitre saw if available.

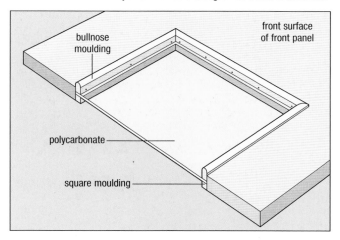

10 **OPTION:** Fit a viewing panel to the front of each unit to show its contents. Select an area on the front panel between two shelves and make a rectangular cutout with a jigsaw. Glue and pin short lengths of **bullnose** moulding to the front edges, mitring the corners as in step 8. Cut a piece of 3mm (⅛") polycarbonate to fit the aperture and secure with square moulding.

Hallway shelving

In tackling this project you can have the best of both worlds: a smart array of shelving to make maximum use of the height available under the staircase, and a small cupboard at the foot of the stairs for more utilitarian items. The width of a standard staircase can easily accommodate a shallow bookshelf without making the doorway too small.

The cupboard and door are simple to make using MDF or other sheet material, fixed to a **stud partition** and painted to match the staircase; as a finishing touch, glue and pin lengths of **panel moulding** to the door and side panel, mitred at the corners.

To make the shelves, lengths of solid timber are recommended as they can support a heavier load without sagging. Square-edged pine floorboards are ideal, and can be made quite presentable with a coat of varnish and brass shelf supports.

Take extra care in this project to fix the door frame and shelf uprights absolutely vertical, and use a spacer to build the units a constant width. This will ensure that all the shelves are fully interchangeable.

By keeping books of the same height together, no space is wasted.

MEASURING UP

First decide on the number of shelf units you require based around a standard width: between 600 and 750mm (24 and 30") is suitable for a bookshelf. The remainder of the space available can then be taken up by making the cupboard to suit.

- Cut the bases for the units 3mm (⅛") longer than the shelves, and use as spacers to plot the positions of the shelf uprights at floor level
- Use a plumb line or spirit level to transfer these positions to the underside of the stair string, and measure the height of each upright
- Make a template of the stair angle with an offcut of plywood or hardboard, and use this to mark out the side panel of the cupboard and the top panels of the bookcase

MATERIALS

- Planed softwood:
  150x22mm (6x⅞"):
  shelves, uprights, bases,
  top panel
  75x25mm (3x1"): plinths
  25x25mm (1x1"): battens for
  fixing bases

- Brass shelf supports and sockets

OPTIONAL:
- Angle brackets to secure uprights

TOOLS

- Adjustable bevel gauge ●
- Block plane or surform
- Combination square
- Drill bits for pilot holes and
- to suit shelf sockets
- Panel saw or jigsaw ●
- Plumb line & bob
- Screwdriver

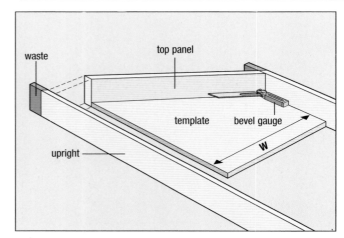

1 Cut the uprights to their respective lengths, and mark the angle required at the top with a bevel gauge. Use a block plane to remove the waste and ensure a neat fit to the stair angle. Do the same for the top panels, using the template you have made to check for an accurate fit. Let the width (W) of the template match the bases, to help you keep the unit square during assembly.

Assembling the hallway shelving

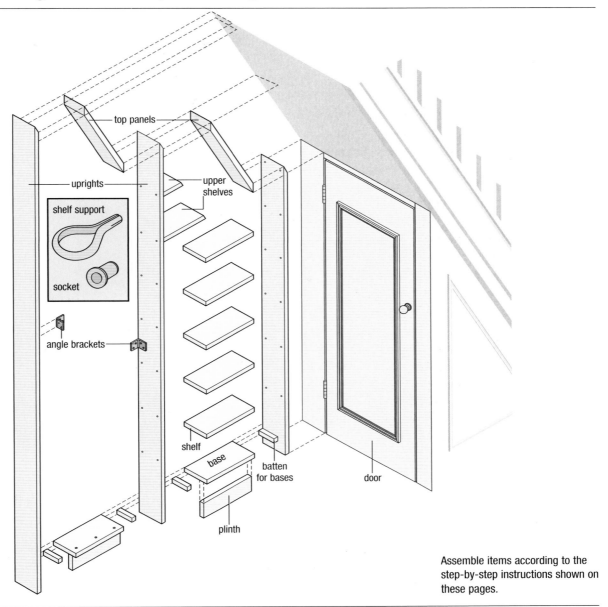

top panels

uprights

upper shelves

shelf support

socket

angle brackets

shelf

base

batten for bases

door

plinth

Assemble items according to the step-by-step instructions shown on these pages.

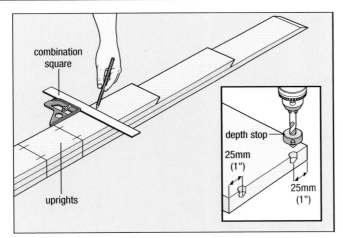

combination square

depth stop

25mm (1")

25mm (1")

uprights

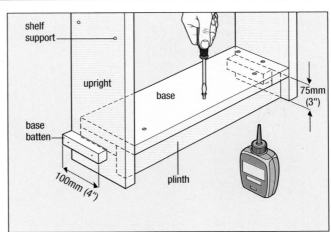

shelf support

upright

base

base batten

100mm (4")

plinth

75mm (3")

2 ★ Before proceeding, pre-drill all the holes needed for the shelf sockets. Lay the uprights together, flush at the bottom ends, and use a combination square to mark out the socket positions, 25mm (1") from the edges. Drill the holes using a depth stop on the drill bit to the required depth *(see inset)*. Finally fix battens to the uprights to locate the bases, 75mm (3") from the bottom edge.

3 Glue and screw the units together, fixing the top panels and bases, and add the plinths. Where the upper shelves meet the sloping top panels, screw up at an angle through each shelf to lock the unit together. Fix the assembly to the rear wall with angle brackets, maintaining a constant width for the shelves. Then cut all the shelves to length, using one as a pattern, and sand edges smooth.

Understairs cupboard

It could be said that the appeal of this project hinges on the ingenuity of the design for the door under the stairs. Whether your staircase is a straight flight, as in the previous project, or a double flight with a half-landing as in this picture, the problem is the same: a narrow cupboard with restricted access and a 'black hole' at the back of your storage space. The solution is to construct a special door which exposes two sides of the cupboard, allowing plenty of light and access for quite bulky items. A further advantage is that, in the open position, the door takes up less space in the hallway.

Construction of the door is not as complicated as it seems: it can be made of standard MDF or other sheet material, and hung in the normal way before shaping the side panel to suit the angle of your stairs. One important note: if the top landing of your staircase is supported by a newel post at the front corner of the cupboard, take expert advice: do not remove it if it is structural.

Two levels of access are provided by the triangular and full doors.

MEASURING UP

Whether you are working to an existing door frame under the stairs, or have built a new partition for the purpose, the method is much the same. The crucial part of the project is fitting the angled side panel. For an accurate fit, make a cardboard template to fit the angle under the stairs, and slide it into position.
- Make the template oversize at the front, and use a straight edge on the face of the partition to mark the cut line
- Use a spacing block 150mm (6") wide to transfer the angles of the upper string and lower handrail to your template
- Cut out the triangular shape, making a smooth curve where the two lines meet, and copy the finished template to a sheet of MDF

TOOLS

- Basic Toolkit
- Drill bits for pilot holes
- G-cramps
- Jigsaw
- Spirit level

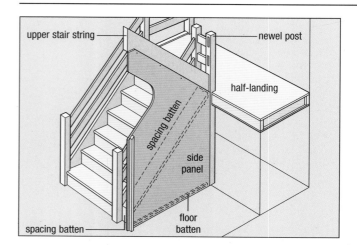

MATERIALS

- MDF 19mm (¾"): side panel, door panel and door side panel
- Planed softwood:
  150x25mm (6x1"): spacing battens
  75x25mm (3x1"): door jamb, doorstop
- 25x25mm (1x1"): doorstop batten, floor batten
- Lay-on hinges, available from DIY stores or mail-order catalogues:
  5 x unsprung for door panel
  3 x sprung for door side panel

1 Cut out the side panel using the template. (Use the triangular offcut to make the door side panel.) Screw to the inside of the upper stair string at the top, and fit a floor batten to locate the bottom edge. Cut spacing battens to fill the gap between the side panel and the lower stair string. Clamp and screw through from the inside of the side panel.

Assembling the understairs cupboard

Assemble items according to the
step-by-step instructions shown on
these pages.

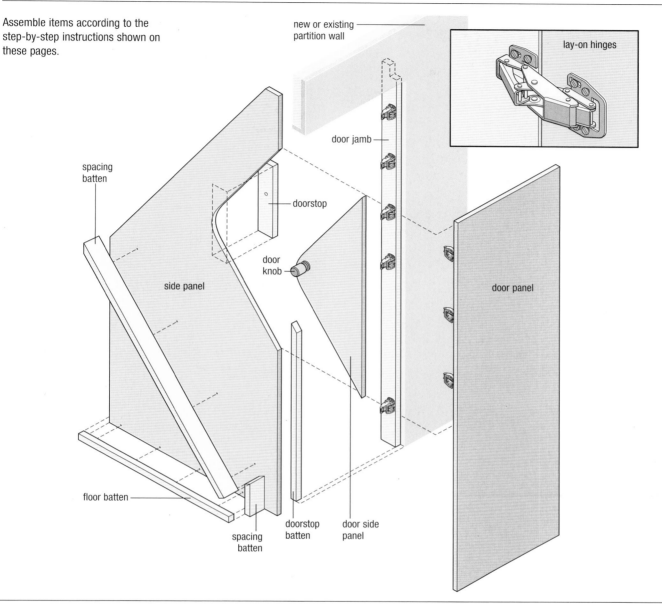

new or existing
partition wall

lay-on hinges

door jamb

spacing
batten

doorstop

door
knob

side panel

door panel

floor batten

spacing
batten

doorstop
batten

door side
panel

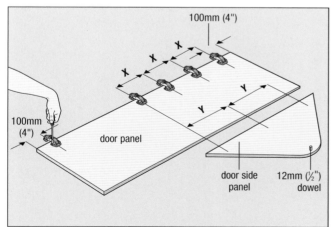

100mm (4")

100mm
(4")

door panel

door side
panel

12mm (½")
dowel

2 Measure from the inside edge of the side panel to the other side
of the door opening and cut the door to size. Allow 3mm (⅛") all
round for clearance. Fit a **door jamb** to receive the hinges,
notched at the top for extra support. Fit hinges to the door panel and
equalise the hinge positions, as shown, to support the extra weight of
the door side panel.

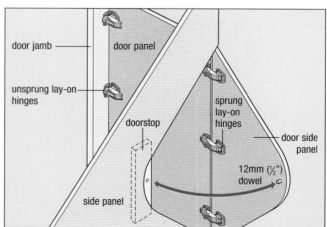

door jamb

door panel

unsprung lay-on
hinges

sprung
lay-on
hinges

doorstop

door side
panel

side panel

12mm (½")
dowel

3 Offer up the door side panel with the door in a closed position
and smooth the edges for a snug fit. Whereas the door itself is
hung on unsprung hinges, here you can use the spring-loaded
version which will keep the side panel snapped closed when required.
A catch is not really necessary: just fit a small **dowel** to the panel and
drill a matching hole in the doorstop.

PROJECT

3

UNDERSTAIRS CUPBOARD

THE SPACE SAVER BOOK

77

Instant storage ideas for hallways

The hallway is a good place to store things that need to be kept, but are not in constant use. Good organization is the keynote in that case, with separate containers for different kinds of items so that they do not get muddled up when one of them is needed. A similarly well-planned approach also allows extra space to be used to store items in regular use without the hall appearing untidy.

Cardboard boxes with lids or drawers are ideal for filing important, but rarely needed, papers, such as instructions for household equipment, and for 'archiving' correspondence or even accounts. The drawers can also be used for easily lost items, such as scissors and string.

Airtight tins in a range of different sizes are, of course, widely used for storing perishable foods, but they can also be used as containers, perhaps stacked under the stairs, for such things as screws and nails.

A lightweight, aluminium trolley can be used to extend kitchen storage space. Castors make it easy to move into the kitchen when the contents are required or out of the way for cleaning the floor in the confined space under the stairs.

A system of sliding wire baskets, available in a range of sizes, is very versatile and extremely easy to install and use.

Glossary of terms and materials

The terms listed here appear in **bold** *type in the text of this book. See also* Toolkit *on pages 6-7 for additional information and illustrations of the items marked with an asterisk * below.*

TERMS

architrave lengths of timber with moulded profile surrounding a doorframe

bench plane* large plane, used two-handed, for achieving a good true and straight edge on a panel or length of timber.

block plane* small plane which can be used one-handed and is ideal for fine finishing and smoothing

bradawl* small hand-held instrument with sharp point for marking or starting a screw hole

brace section of timber, usually glued and screwed to a panel, to add strength or rigidity to the construction

bullnose half-round moulding, as used for finishing the front edge of a stair-tread

carcass the basic box assembly of any construction, such as a cupboard or shelving unit

chamfer to remove the sharp corner of a section of timber and produce a smooth bevelled edge

clearance hole hole drilled through a panel to allow clearance for the shank of a woodscrew

combination square* square with sliding adjustable blade, for marking out and gauging work as well as checking square corners and marking mitres

connecting bush threaded nylon fitting, inserted flush with the surface, which receives a connecting bolt

cornice concave or moulded timber section, typically used at the junction of walls and ceiling

cross dowels threaded metal rod inserted across the grain of a timber member *(see connecting bush)*

door blank oversize flush door panel, typically 2100x900mm (7x3'), for cutting to suit a special size

door jamb vertical component of a door-frame, which receives the hinges or latch

dowel* cylindrical length of wood used for connecting a butt joint, sometimes fluted to allow good glue adhesion

dowel centre point brass insert, with a central point, used to mark the matching hole positions of a dowel joint

dowel joint butt joint between two sections of timber strengthened with wooden dowels

edging clamp* specialised G-cramp with an additional screw at right angles to the first, used for applying edge trim

former battens radial battens, forming a shape around which a thin plywood panel can be bent to form a curve

glazing tape adhesive fabric tape for protecting and cushioning the edges of glass when fixed with timber moulding

halving joint joint where two timber members intersect, each one cut away by half its thickness to receive the other

handsaws

    coping saw* fine toothed saw with narrow blade for cutting curves

    hacksaw* saw with high tensile steel blade for metal cutting

    panel saw* the standard saw for cutting or ripping large panels to size

    tenon saw* essential tool for cutting simple joints and mouldings

hinge boring bit* special drill bit for boring holes suitable for concealed hinges available from the same suppliers as the hinges

hockey-stick moulding used to trim the edges of a panel, with a J-shaped rounded end in the profile of a hockey-stick

housing joint shallow groove in a panel or cross member, housing the end of another panel or timber component

joint blocks plastic corner blocks for easy connection of two components at right angles

laminated glass two thin layers of glass bonded by a thin plastic film, making a safe, shatter-proof glazed panel

marking gauge* cabinet-maker's tool, for gauging the depth of a joint with accuracy and consistency

mitre halving the angle of a corner joint where two members intersect in a right angle, two 45 degree angles

modular describes a design method based around simple standard-sized components combined to form a more complex construction

panel connector two semi-circular fittings connected by a bolt, for edge-jointing two panels

panel moulding small decorative moulding, usually of triangular section, for locating or trimming round a panel

pilot hole hole drilled to receive the threaded end of a woodscrew, to assist location and prevent the wood splitting

profile cutter special cutter fitted to a router, used to follow a template for reproducing a particular profile

profile gauge* handy tool like a comb with sliding teeth, which can be pressed against a moulding to copy the profile

quadrant moulding small moulding in the profile of a quarter circle, used for decoration or trimming the edge of a panel

rebate to cut a rectangular, stepped recess along the edge of a section of timber

scribe to mark or shape the end of a timber section to fit around a moulding or profile for a neat joint

spacing rod batten or strip of wood, cut to a fixed length

and used as a reference for defining or marking out the work

stud partition hollow dividing wall, made by cladding a simple framework of lengths of timber (studs) nailed together

surform* alternative to a plane, with many small teeth, good for roughing out a shape before final finishing

MATERIALS

chipboard small wood particles compressed in a flat hard panel with a fibrous core, used for flooring or furniture components

furniture board chipboard panels, faced with melamine film or wood veneer, ideal for carcass construction

hardboard fibrous sheet material, usually 3mm (⅛") thick, used for non-structural panels or templates

hardwood more expensive but better quality timber, from broad-leaved trees, ideal for fine finishing or joinery

MDF medium density fibreboard – all-purpose sheet material, available in a range of thicknesses with a smooth sanded surface, suitable for home woodworking projects of all kinds

pineboard laminated pine strips, factory-glued and sanded to make wide flat boards ideal for furniture construction

plywood thin layers of wood, glued alternately in sheet form to make a strong building board, ideal for shelving or general carcass construction

softwood inexpensive, general purpose timber, from pine or other coniferous trees. Obtainable in standard sizes as planed (smooth and square edged) for accurate constructional work, or rough sawn for studwork

veneer thin layers, usually of hardwood, glued to the face of MDF or plywood panels for decorative purposes

INDEX *(The terms listed in the glossary on page 79 are not repeated in this index)*

ACKNOWLEDGEMENTS

Key TL=top left; TC=top centre; TR=top right; CL=centre left; C=centre; CR=centre right; B=bottom; BL=bottom left; BC=bottom centre; BR=bottom right

Elizabeth Whiting Associates
8, 9, 10, 14, 16,19TR, 20, 21, 22, 26, 28, 32, 33, 34, 38, 40, 44, 45, 46, 50, 52, 56, 57, 58, 62, 64, 68, 69 and 70 (Designer:

Christian Stocker), 74, 76,
**Elizabeth Whiting Associates/
David Giles** 30TL, TR, BL, 31TR, 66TL, TR, BL, 67BL, BR, 78TR
Houses & Interiors 19BL,
**Houses & Interiors/Gwenan
Murphy** 19BR

**The Cotswold Company •
01276 606 029**
42TL, TC, C, 54TL, BR, 55TL, BL,

Elfa (Storage Systems) •
  01844 292 856
1, 18, 19TL, 42BL, 67TL, 78BR
Homebase • 0645 801 800
43TR, 78BL
IKEA Ltd • 0181 208 5607
55TR,
The John Lewis Partnership
*For details of your nearest branch
  phone* **0171 629 7711**
30BL, 42CR, 43TL, B, 66BR,

Muji • 0171 323 2287/2285
31TL, BL, BR, 54TR, 55BR, 67TR, 78TL
Front cover:
Elizabeth Whiting Associates
  (main pictures)
Tools courtesy of Black & Decker
  and Stanley®
Back cover:
Author's portrait: Steven Differ
Illustrations: Andrew Green